American Institute of Chemical Engineers

Fuels and Petrochemicals Division Conference

Presentations at the
2008 AIChE Spring National Meeting

April 6-10, 2008
New Orleans, Louisiana, USA

Printed from e-media with permission by:

Curran Associates, Inc.
57 Morehouse Lane
Red Hook, NY 12571
www.proceedings.com

ISBN: 978-1-60560-221-9

Some format issues inherent in the e-media version may also appear in this print version.

ISBN: 978-1-60560-221-9

American Institute of Chemical Engineers
Proceedings
Three Park Avenue
New York, NY 10016-5991
Phone: 212-591-8100

www.aiche.org

American Institute of Chemical Engineers

Fuels and Petrochemicals Division Conference
2008

TABLE OF CONTENTS

143a 1-Octene from Butadiene – New Technology Based on Heterogeneous Catalysts

Andrzej Krzywicki, Xiaoliang Gao, Oleksiy Golovchenko, Stacy Johnston, and Antonio Nicola.
NOVA Chemicals Corp., 2928 - 16th Street NE, Calgary, AB T2E 7K7, Canada

1-octene has been used as one of the basic comonomers in the production of polyethylene. Its current availability is limited. At this time, there is no proven commercial process that allows making 1-octene only with high conversion and high selectivity. The main source of this comonomer is a product mixture from oligomerization of ethylene resulting from Linear Alpha Olefin (LAO) process. The LAO processes generate a broad range of olefininc hydrocarbons. The main drawback of this process is in the production of a mixture of products with a limited possibility of influencing its composition. Recently, Sasol claimed the development of selective ethylene tetramerization technology (1-3), and Dow Chemical started up installation converting butadiene into 1-octene. There is only scarce information about these developments.

There were several attempts made by academia and industrial companies to develop a process based on known chemistry - telomerization of butadiene with telomers such as: water, alcohols, or organic acids. The 1-octene synthesis from butadiene is a three-step process. First, and the most difficult step is the dimerization of butadiene connected with the addition of water (telomerization) to make 2,7-octadienol (reaction 1), or with the addition of methanol to make 1-methoxy-octadiene-2,7 (reaction 2).

1. $2\ CH_2{=}CH{-}CH{=}CH_2 + H_2O \rightarrow CH_2{=}CH{-}CH_2{-}CH_2{-}CH_2{-}CH{=}CH{-}CH_2{-}OH$

2. $2\ CH_2{=}CH{-}CH{=}CH_2 + CH_3OH \rightarrow CH_2{=}CH{-}CH_2{-}CH_2{-}CH_2{-}CH{=}CH{-}CH_2{-}OCH_3$

During this step, it is critical to have high selectivity catalyst to avoid the formation of higher oligomers of butadiene and isomeric diene derivatives. Catalyst, nature of electron donor, and reaction medium are critical parameters for this stage. Several patents have been granted to Dow Chemical (4), Kuraray (5), Oxeno (6,7), and Shell (8).

In the second step (reaction 3), the unsaturated ether or alcohol is hydrogenated to 1-substituted octane. This reaction uses classical hydrogenation catalysts (e.g. nickel, palladium, or platinum on alumina) working effectively under mild reaction conditions.

3. $CH_2{=}CH{-}CH_2{-}CH_2{-}CH_2{-}CH{=}CH{-}CH_2{-}O{-}CH_3 \rightarrow CH_3{-}(CH_2)_6{-}CH_2{-}O{-}CH_3$

The last stage is cracking of 1-methoxy-octane (1-MOAN) to 1-octene and methanol (reaction 4). A very selective catalyst is required to avoid the migration of double bond along the carbon chain.

4. $CH_3\text{-}(CH_2)_6\text{-}CH_2\text{-}O\text{-}CH_3 \rightarrow CH_3\text{-}(CH_2)_5\text{-}CH{=}CH_2 + CH_3OH$

According to the patent literature several catalysts based on gamma-alumina were claimed. Alumina modified with alkali metal ions, looked like an especially good catalytic system for cracking (5-7) ethers or alcohols.

Telomerization

Our approach to the development of a new heterogeneous process for the manufacturing of 1-octene from butadiene was focused on the use of solid catalysts to avoid costly procedures associated with the recovery of homogeneous catalysts, and handling large volumes of organic solvents. For the telomerization step we developed palladium containing catalysts active and selective in telomerization of butadiene with methanol. During the preparation and activation of the catalysts, it is essential to maintain palladium atoms in an ionic form. Therefore catalysts need to be activated in an oxidizing atmosphere at elevated temperature (450-850°C). Reduction of palladium ions by hydrogen or by high temperature treatment caused total loss of catalytic activity in telomerization reaction.

These catalysts have been showing very good stability and selectivity toward 1-methoxy-octadiene-2,7 (1-MODE), exceeding 95% in the mixture with 3-methoxy-octadiene-1,7 (3-MODE) (11). These results were achieved in a fixed bed reactor system under continuous flow of reactants. Residence time of 1 hour was significantly shorter than that reported for batch reactor systems of 6-18 hours (9,10). At 85°C under the pressure of 1.35 to 2.0 Mpa, we achieved between 60 to 85% conversion of butadiene, with the overall 1-MODE yields above 60%. The main by-products of the telomerization step were oligomers of butadiene (octatriene and vinyl-cyclohexene). Developed catalysts for telomerization demonstrated good stability, not changing activity over several hours of operations. In addition, these catalysts could be on-line regenerated without loosing activity and selectivity.

Hydrogenation

Hydrogenation of separated from telomerization step products 1-MODE occurs readily over either Pd or Ni catalysts. Seven commercially available catalysts were tested and compared under similar conditions. Five Pd catalysts compared under isothermal conditions gave a range of activity and selectivity towards the desired product, 1-methoxyoctane (1-MOAN). Two Ni catalysts were compared under adiabatic conditions.

In our study, Ni catalysts showed higher activity compared to Pd catalysts under similar conditions. This was clearly shown as a correlation between conversion and flow rate/residence time. Nickel catalysts maintained minimum requirements for activity and selectivity at much higher space velocities compared to Pd catalysts. Nickel containing catalysts demonstrated also longer lifetime.

Cracking of 1-methoxy-octane

For the third step, the cracking of the 1-MOAN to 1-octene is more problematic as the produced 1-octene tends to isomerize to internal octenes over solid oxide catalyst. Suppressing the isomerization of 1-octene to the minimum has been a focal point of our efforts. Several commercially available alumina catalysts were tested. The best results were achieved in the presence of a large pore alumina having an average pore diameter larger than 150 Angstroms (12). Using this catalyst, we achieved in the temperature range of 280-330°C, conversions above 65%, and very high selectivities of 1-octene formation exceeding 96%. The large pore alumina maintained its activity and selectivity for a long period of time. Long-term runs showed only small decline in catalyst performance over a period of 800 hours. This high selectivity allows using the 1-octene fraction directly as a comonomer for ethylene polymerization. With the increase of cracking temperature there was a slow decline in the selectivity of 1-octene. The selectivity could be improved by selective blockage of acid sites by the addition of small amounts of ammonia (13).

The development of technology on lab scale was not fully successful in recovering methanol according to reaction (4) above. Cracking products contained di-methyl ether as the main component.

Bibliography

1. PCT Pat. Appl. WO 2004/056478, WO 2004/056479, and WO 2004/056480; July 8, 2004; Sasol Technology Ltd.

2. M.J. Overett, K. Blann, A. Bollmann, J.T. Dixon, F. Hess, E. Killian, H. Maumela, D.H. Morgan, A. Neveling, S. Otto; Chem. Commun., 2005, 622-624.

3. M.J. Overett, K. Blann, A. Bollmann, J.T. Dixon, D. Haasbroek, E. Killian, H. Maumela, D.S. McGuinness, D.H. Morgan; J.Am.Chem.Soc.; 127, 10723-30 (2005).

4. US Pat. 5,254,782; October 19, 1993; Dow Chemical Co.

5. US Pat. Appl. 20060111594; May 25, 2006; Kuraray Co., Ltd.

6. US Pat. Appl. 2004/0059170; March 25, 2004; Oxeno Olefinchemie Gmbh

7. PCT Pat. Appl. WO 03/031379; April 17, 2003; Oxeno Olefinchemie Gmbh

8. Can. Pat. 2,010,325; January 25, 2000; Shell Canada

9. Benvenutti F., Carlini C., Raspolli Galetti A.M., Sbrana G., Marchionna M., Patrini R.; J.Mol.Catal., 137 (1999), 49

10. Lee B.I., Lee K.H., Lee J.S.; J.Mol.Catal., 156 (2000), 283

11. US Pat. Appl. 20070112238, May 17, 2007; NOVA Chemicals S.A.

12. US Pat. Appl. 20070203381, August 30, 2007; NOVA Chemicals S.A.

13. US Pat. Appl. 20070100186, May 3, 2007; NOVA Chemicals S.A.

THERMODYNAMIC PREDICTION OF THE OLEFIN PRODUCT DISTRIBUTION IN THE FISCHER-TROPSCH SYNTHESIS

Cornelius M. Masuku, Diane Hildebrandt, and David Glasser; University of the Witwatersrand, Johannesburg, South Africa

Abstract

A substantial number of papers postulate various mechanisms for chain growth in Fischer-Tropsch synthesis. Most agree that the synthesis mechanism may be divided into a number of steps: (a) Adsorption of reactants, (b) chain initiation, (c) chain growth, (d) chain termination, (e) desorption of products, and (f) readsorption and further reaction. It was postulated in 1926 that the synthesis involved oxygenated intermediates. This proposal was later extended in 1951 to a mechanism that explained many characteristics of synthesis products, such as carbon number and isomer distributions. Here the intermediate was postulated to be similar to an alcohol. The incorporation of radioactive alcohols into the synthesis largely confirmed these predictions.

In 1953 some concluded that primary alcohols, aldehydes, and 1-olefins were the primary synthesis products. Apparently this conclusion demanded no major revision of the concept of oxygenated intermediates in the synthesis. Further developments showed that α-olefins are primary products and that paraffins are formed from them.

As a primary assumption we suggest that the product distribution may be determined by Equilibrium considerations. Thus this work uses thermodynamics to predict the equilibrium distribution assuming that α-olefins are the primary products. The results obtained show that α-olefins follow a single alpha product distribution at Fischer-Tropsch reaction conditions which is consistent with many experimental results. However, assuming that the alcohols and parraffins are the primary products is not consistent with experimental results. This further supports that α-olefins are the primary products of the FT synthesis and paraffins are formed from them since the whole product distribution is determined by olefins.

Keywords:

Fischer-Tropsch Reaction, α-olefins Primary products, FT Thermodynamic Equilibrium, Product Distribution Model, Chain propagation probability (α), ASF.

Introduction

Reactions of carbon monoxide with hydrogen at various temperatures and pressures over such different catalysts as iron, cobalt, nickel, ruthenium, zinc oxide, and thoria are unlikely to proceed by identical mechanisms. On the other hand, the mechanisms of the several reactions operative in the hydrocarbon synthesis process probably have some valid extensions to related processes. It is a question whether differences in product composition associated with different catalyst or conditions reflect different mechanisms or merely reflect changes in the relative rates of competing reactions. An important aspect of the problem is the determination of the primary and secondary products and the consideration of the secondary conversions which the primary products may undergo.

Olefins and alcohols are mentioned frequently as likely primary products of synthesis with either cobalt or iron catalysts.

Reaction Mechanism

A substantial number of papers postulate various mechanisms for chain growth in Fischer-Tropsch synthesis. The Synthesis mechanism may be divided into a number of steps: (a) Adsorption of reactants, (b) chain initiation, (c) chain growth, (d) chain termination, (e) desorption of products, and (f) readsorption and further reaction [1]. Elvins and Nash in 1926 postulated that the synthesis involved oxygenated intermediates. This proposal was extended by Storch, Golumbic, and Anderson [2] in 1951 to a mechanism that explained many characteristics of synthesis products, such as carbon number and isomer distributions. Here the intermediate was postulated to be similar to an alcohol. The incorporation of radioactive alcohols into the synthesis has largely confirmed these predictions.

In 1953 Weitkamp and Frye [3] concluded that primary alcohols, aldehydes, and 1-olefins were the primary synthesis products. Apparently this conclusion demands no major revision of the concept of oxygenated intermediates in the synthesis. Further developments showed that α-olefins are primary products and that paraffins are formed from them [4, 5].

This work uses thermodynamics to minimise the Gibbs free energy for α-olefins to determine the equilibrium product distribution.

Primary Products

Alcohols, olefins, and paraffins have been suggested as possible primary products of the Fischer-Tropsch process. Aldehydes, ketones, acids, and aromatic hydrocarbons are present and must also be considered. Two techniques for the identification of primary products are available [3]. One technique is to find the products whose relative yields do not approach zero as the conversion of carbon monoxide is progressively reduced. This method is of limited applicability to heterogeneous reaction systems because mass transfer of reactants to the catalyst or of products from the catalyst may be a limiting factor at high space velocities.

The other technique is to find whether possible secondary reactions are operative and, from known thermodynamic equilibria, to determine which classes of components are reactants and which are products.

Equilibrium calculations [3] show that possible secondary reactions involving hydrocarbons are far from equilibrium. In this way it has been established that 1-olefins are partially converted, in secondary reactions, to paraffins and 2-olefins. Reactions involving oxygenated compounds, except for the decarboxylation of acids to ketones, are so close to equilibrium that a choice between alcohols, aldehydes, and acids as primary products could not be made.

Accepted FT Reaction Scheme

The FT synthesis is a surface polymerization reaction [6]. The accepted model is that reactants, CO and H_2, adsorb and dissociate at the surface of the catalyst and react to form chain initiator (CH_3), methylene (CH_2) monomer and water. The hydrocarbons are formed by CH_2 insertion into metal-alkyl bonds and subsequent dehydrogenation or hydrogenation to an α -olefin or paraffin, respectively. According to Anderson [7], the product distribution of hydrocarbons can be described by the Anderson-Schulz-Flory (ASF) equation:

$$m_n = (1-\alpha)\,\alpha^{n-1}$$

In which m_n is the mole fraction of a hydrocarbon with chain length n and the growth probability factor α is independent of n. α determines the total carbon number distribution of the FT products. Hence, a plot of the logarithm of the molar concentration versus the carbon number would produce a straight line plot whose slope is related to alpha (α), the chain propagation probability. The total product yield decreases exponentially with increasing chain length.

Most studies assume the Anderson-Schulz-Flory (ASF) distribution to model the product selectivity. The ASF model describes the entire product range by a single parameter, α, the probability of the addition of a carbon intermediate to a chain.

Model Development

As a primary assumption we suggest that the product distribution may be determined by Equilibrium considerations. This work considers a feed to a reactor of 1 and 2 mol for CO and H_2 respectively to work out equilibrium distribution for various FT products. For simplicity of simulation 30 linear paraffins and up to C20 α-olefins are considered. The conditions used for calculations are 8 atm and 200 °C except where stated otherwise.

Equilibrium Conversion Modelling

Considering Paraffins as Primary Products

The Equilibrium Product Distribution results for considering Paraffins as Primary products for a feed of 1 and 2 mol for CO and H_2 respectively at 8 atm and 200 °C are presented in Figure 1 below.

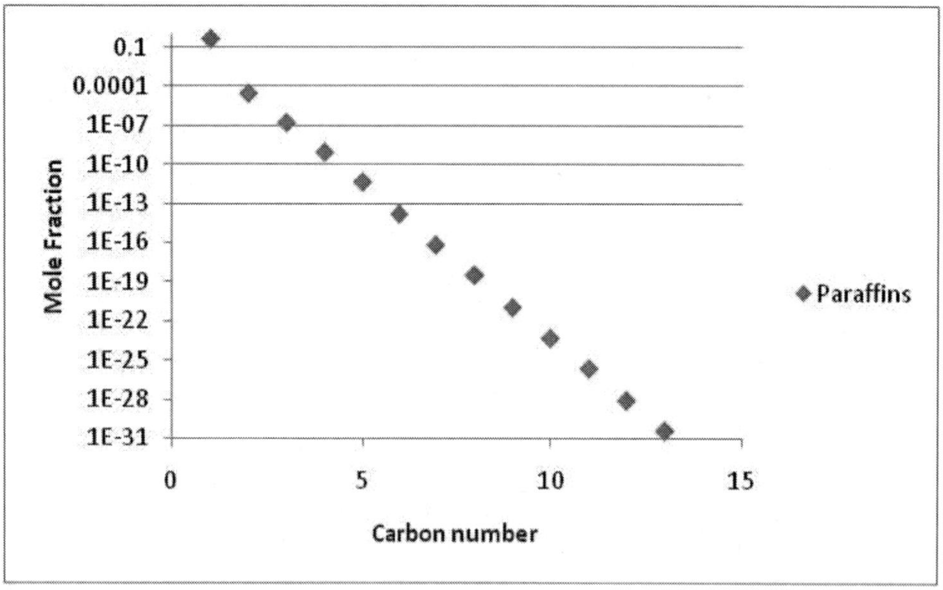

Figure 1: Equilibrium Product Distribution for Paraffins as Primary Products.

The Equilibrium results presented in Figure 1 show that almost exclusively methane (an alpha of ≈ 0.004) should be formed from CO and H_2. This is not observed in practice.

Considering both Paraffins and Olefins as Primary Products

Since considering only Paraffins does not describe the observed product distribution both Paraffins and Olefins were considered and the results obtained are presented in Figure 2 below.

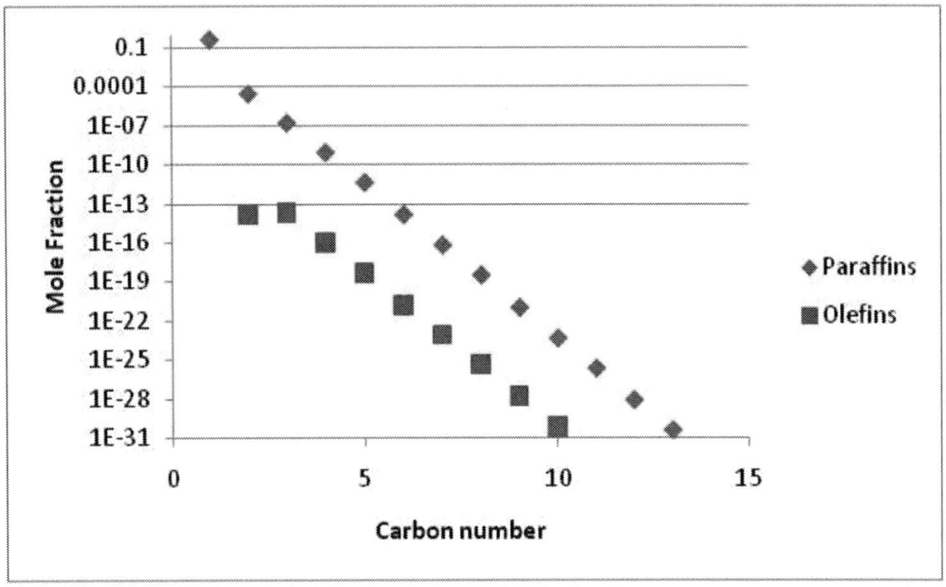

Figure 2: Equilibrium Product Distribution for Paraffins and Olefins as Primary Products.

Figure 2 show that considering both Paraffins and Olefins still produces almost exclusively methane (alpha as before) which is inconsistent with observed product distribution.

Considering both Paraffins and Olefins excluding methane as Primary Products

Suppose we assume that methane reaction is not at equilibrium, and methane was excluded from equilibrium considerations, the results obtained are as presented in Figure 3 below.

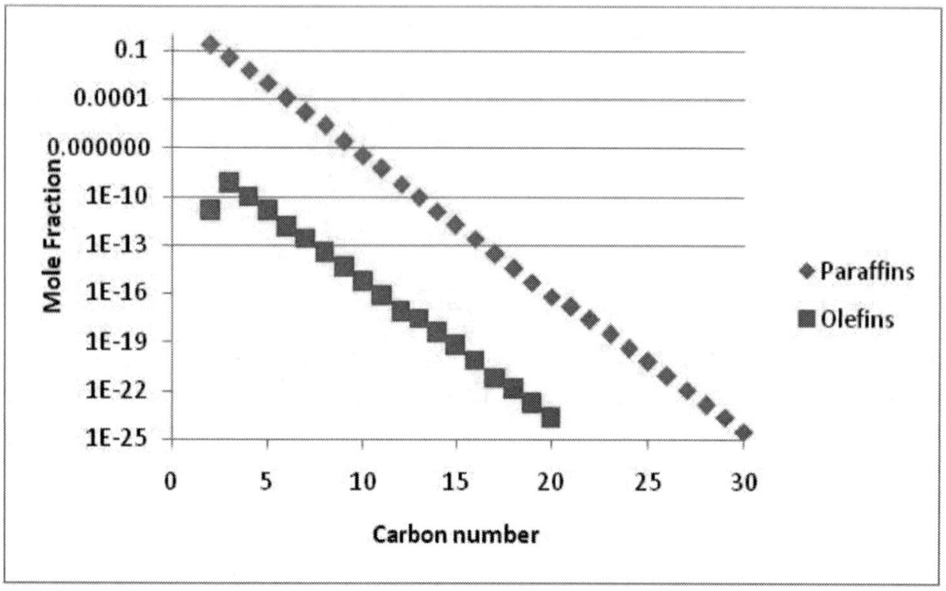

Figure 3: Equilibrium Product Distribution for Paraffins and Olefins excluding methane as Primary Products.

The results in Figure 3 show an ASF product distribution for both Paraffins and Olefins however the predicted alpha is 0.1 for both species. This implies that if equilibrium existed for both Paraffins and Olefins a low-alpha (α = 0.1) would be expected.

Considering Olefins as Primary Products

Olefins had been extensively proposed to be the primary products for the FT Reaction and were therefore considered as primary products. The results are shown in Figure 4 at different temperatures.

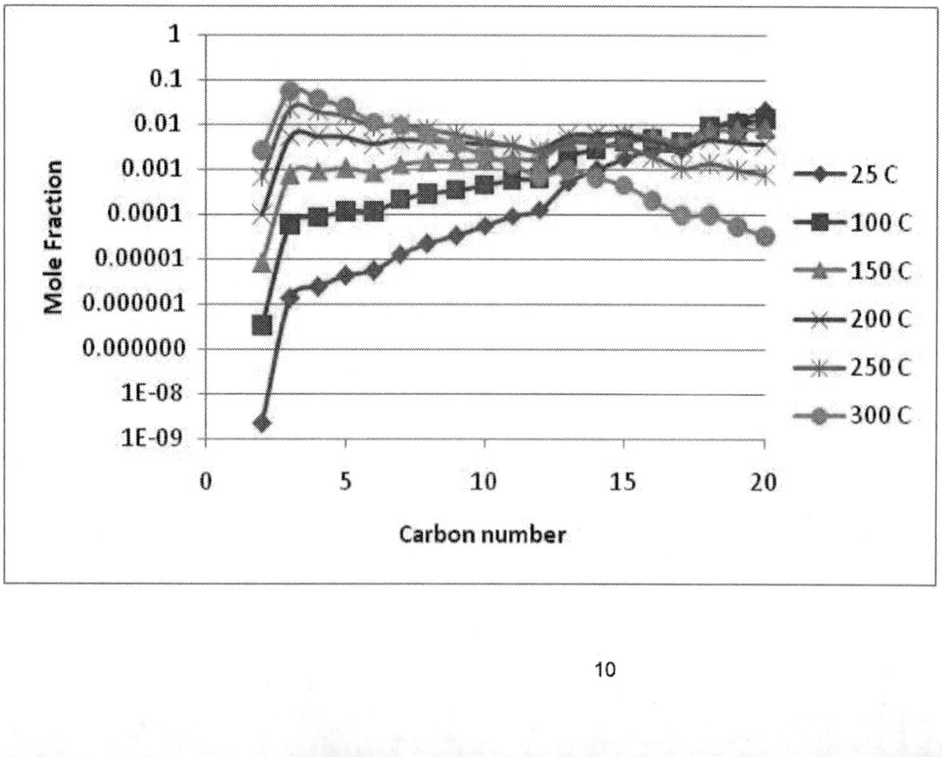

Figure 4: Equilibrium Product Distribution for Olefins as Primary Products at different Temperatures.

It is interesting to note that under standard conditions (1 atm, 25 °C) Olefins show a polymerisation reaction in which the chain would grow indefinitely. This is not the case in the normal FT reaction where the chain-growth decreases exponentially with chain-length. The results obtained at other temperatures are also presented in Figure 4. It can be seen that the slope decreases with increasing temperature.

At 200 °C the predicted alpha is one ($\alpha = 1$) and that is the temperature at which appreciable FT products are obtained. Alpha decreases to 0.67 at 300 °C. These results are consistent with the range of alphas reported experimentally and industrially. These results support the single alpha ASF product distribution and suggest that olefins are the primary FT products and paraffins are formed from them.

Conclusion

The results obtained show that overall thermodynamic equilibrium does not exist. They do however suggest that the olefin product distribution might be equilibrium determined. This also shows that at equilibrium α-olefins follow a single alpha product distribution under Fischer-Tropsch reaction conditions. This study suggests that olefins might be primary FT products and paraffins are formed from them. It also suggests that methane might be formed by a separate reaction which is kinetically determined.

References

1. Anderson R. B., Kinetics and Reaction Mechanism of the Fischer-Tropsch Synthesis, *Catalysis*, **1956**, *4*, Reinhold New York, 335.

2. Storch H. H.; Golumbic N.; Anderson R. B., The Fischer-Tropsch and Related Syntheses, New York; John Wiley & Sons; Inc., **1951**.

3. Weitkamp A. W.; Frye C. G., Relation of product composition to reaction mechanism, *Ind. Eng. Chem.*, **1953**, *45*, 363-367.

4. Satterfield C. N.; Huff G. A. Jr., Carbon number Distribution of Fischer-Tropsch Products Formed on an Iron Catalyst in a Slurry Reactor, *J. Catal.*, **1982**, *73*, 187-197.

5. Sudheimer G.; Gaube J., Fischer-Tropsch Synthesis: Kinetic Investigation and Process Design for Straight-chain 1-alkenes, Ger. Chem. Eng., **1985**, 8, 195-202.

6. Anderson R. B.; Friedel H. and Scorch H. H., Fischer-Tropsch reaction mechanism involving stepwise growth of carbon chain, *J. Chem. Phys.*, **1951**, *19*, 313-319.

7. Anderson R. B., *Catalysts for the Fischer-Tropsch Synthesis*, Reinhold New York, **1956**, *4*, 337.

Applying Computational Fluid Dynamics (CFD) to Improve the Para-xylene Crystallization Technology

Qingjun Meng, BP Americas Inc., Naperville, IL, USA

Derek Colman, BP Chemical Ltd., Sunbury on Thames, UK

Abstract

In this paper, computational fluid dynamics (CFD) modeling is employed to study the fluid dynamics and heat transfer in BP's PX crystallizers. Due to the low mean velocity through the crystallizer and external cooling, significant buoyancy effects provide a challenge to the modeling. To fully understand the fluid dynamics and its effect on heat transfer in this application, a pilot plant scale vessel and the commercial crystallizer are investigated. For the isothermal pilot plant model, CFD predicted residence time distribution (RTD) compares well with the experimental curve. For the commercial crystallizer, both fluid dynamics and heat transfer are modeled and a simple crystallization model is included to introduce the effect of solids on physical properties and fluid dynamics. Validation against experimental RTD data is reasonable. This model is now being used to help improve the performance of PX crystallizers.

Keywords: CFD, Para-xylene, PX, crystallizers

Introduction

Para-xylene (PX) is the key feedstock of the polyester value chain. Through oxidation and purification PX can be converted to purified terephthalic acid (PTA), which is one of the monomers for poly (ethyl terephthalate) (PET). PET is widely used in the industry as fiber, beverage bottles, etc.

PX is primarily produced from reformate mixed xylenes with roughly 20% PX, 45% meta-xylene (MX), 25% otho-xylene (OX) and 10% ethyl benzene (EB). In BP's proprietary PX technology, PX is recovered from mixed xylenes through eutectic crystallization technology. During more than 40 years of operation and development, this technology has been proved to be energy efficient and forgiving to feedstock quality.

Crystallization is one of the key factors affecting the variable cost of a PX unit because the refrigeration system consumes electricity. By studying mixing and the mechanisms of heat transfer in the crystallizers, it is expected to identify solutions to fundamentally improve the PX crystallization technology as well as the performance of the PX process.

PX crystallization is achieved by lowering the temperature of mixed xylenes to eutectic temperature in scraped wall crystallizers. The difficulty of studying heat transfer in PX crystallizers stems from the complexity caused by the periodic movement of the scrapers and the presence of solid. In a number of published reports, significant simplification has to be made either in the experiments or models [1, 2, 3, 4, 5]. De Goede studied heat transfer in a pilot plant scraped wall crystallizer [6], but the work was more focused on the effect of crystallization on heat transfer while the fluid dynamics in the bulk is ignored.

In this paper, the authors use computational fluid dynamics (CFD) as a tool to study fluid dynamics as well as heat and mass transfer in the PX scraped wall crystallizer. CFD simulations are implemented for both a small scale pilot plant Plexiglas model and a commercial size crystallizer. FLUENT 6.3 is the platform where the models are developed. In the next section, the modeling methodology for both the pilot plant model and the commercial crystallizer will be discussed.

Modeling Methods

Model Setup

The Plexiglas model is scaled down from the commercial crystallizer so in FLUENT the geometries of the two models are almost identical. Some simplification and modification have to be made to implement steady state simulation in the single reference frame (SRF) (Readers are referred to FLUENT 6.3 Documentation [7] for detail about SRF.) due to the presence of rotating parts. Because SRF requires the stationary walls to be a surface of revolution, the outlet pipes of both models are modeled as an annular zone (The reader can consider that in the reference frame of the rotating shaft the annular zone is formed by revolving the outlet pipe about the shaft.). The inlet pipe is modeled as a mass source term and a momentum source term in an annular zone so that the geometry qualifies for the application of the SRF.

In both vessels the flow is mainly driven by the rotating parts, including the supporting arms and scrapers. Therefore care has to be taken to mesh these parts with high resolution so that the momentum equations can be solved with high accuracy around these regions. Given the large volume of the crystallizers the number of meshes is about 3.0 million for the crystallizer model even after the meshes are allowed to grow in regions that are far from the rotating parts. The number of meshes for the Plexiglas model is about 1.6 million. Although there is a symmetry plane in the geometric setup a complete 3D model was used so that any 3D asymmetries that may develop in the flow are not lost.

The turbulence in both the Plexiglas model and the crystallizer is low. The Reynolds number around the scrapers is 10^3 and 10^5 for the Plexiglas model and the crystallizer, respectively, and it is much lower in other regions. To better model the low turbulence flow in the Plexiglas model, the Reynolds Stress Model (RSM) is chosen. By carefully adjusting the under-relaxation factors, the equations converge well. For the crystallizer model, although it is desirable to apply RSM, it was found that the model is extremely difficult to converge. Based on the trials on several other options in FLUENT, Realizable K-epsilon (RKE) model is chosen for the crystallizer model.

The Plexiglas model is run with water at room temperature, therefore there is no heat transfer involved. The flow rate of water is 12 Gallon/hr. For the crystallizers, temperature is specified on the wall and the flow rate is set to the plant value. The shaft speed for both models is the same.

Effect of Crystallization

Crystallization has a significant effect on the flow and heat transfer. During crystallization, PX crystals growing out of the liquid can change the flow field through solid-liquid interactions and buoyancy effects as it triggers the density change. Additionally, a large amount of latent heat (heat of fusion) is released to the surrounding fluid and this can dramatically affect the temperature profile. To reduce the complexity of the model, a simple crystallization model is chosen as,

$$G = k\left(\frac{C}{C_{sat}} - 1\right),$$

where G is the crystal growth rate, k is the crystal growth constant, C and C_{sat} are the concentration and saturated concentration of PX, respectively. This simple relation is easy to implement in the simulation because it can deal with both crystallization and solid melting. When $C > C_{sat}$, G is positive so PX will crystallize from the liquid. When $C < C_{sat}$ and there is solid existing, G is negative so PX solid will melt. In FLUENT, crystal growth is characterized by a source term for the PX solids and a sink term with equal magnitude for PX in the liquid. Melting of solid can be modeled in a similar way. Based on this the latent heat can be calculated by

$$\dot{Q} = G\Delta H_f,$$

where $\dot{Q}$ is the rate of latent heat and ΔH_f is the heat of fusion for PX. Again a heat source term, which could be a sink term, can be specified in the fluid to model either heat release due to crystallization or heat adsorption due to solid melting.

PX solid is modeled as a species in the liquid phase instead of a second phase to reduce computational cost. The approximation is based on the assumption that the solid phase will have minimal effect upon the flow because the solid volume fraction is less than 15% and the particles are small in this particular crystallizer.

The simplification of the crystallization kinetics and the approximation are mainly the result of the tradeoff between the accuracy of the model and the computational cost to converge the model. The validity of the simplifications will be examined by comparing the model results with existing experimental data, such as the residence time distribution (RTD) data presented in the next section.

Results and Discussion

Plexiglas Model

Validation of the CFD model for the Plexiglas model is implemented by comparing the RTD curve obtained in the simulation and that obtained from experiments. In the simulation, once the model is converged, the velocity field is frozen (momentum and turbulence equations are deactivated). Then a pulse of tracer is released from the inlet and the concentration of tracer is monitored at the outlet by solving the species transport equation to obtain the RTD curve.

Based on the comparison of the two curves shown in Figure 1, the CFD model can predict the RTD very well. Although the simulation result overestimates the time it takes the tracer to arrive at the outlet, it predicts the peak and the tail of the RTD curve almost perfectly. The mismatch in the early stage of the RTD curve could result from the modifications that were made to the inlet pipe to apply the SRF algorithm.

Figure 2 shows the velocity profiles on three horizontal planes that cut through the Plexiglas model. In Figure 2 (a), the velocity magnitude contour looks very similar to the rigid body rotation in regions that are away from the rotating scrapers, which indicates that the fluid is swirling as a forced vortex in most of the vessel which by its nature resist radial movement. Consequently Figure 2 (b) shows that although the radial velocity is relatively high around the scrapers, it decays quickly in the center of the vessel. These results are not surprising because it is not expected that the small rotating parts in the pilot plant model can produce much mixing. The animation generated during the tracer simulation also shows that the tracer flows downwards with little dispersion in the radial direction after it enters the vessel.

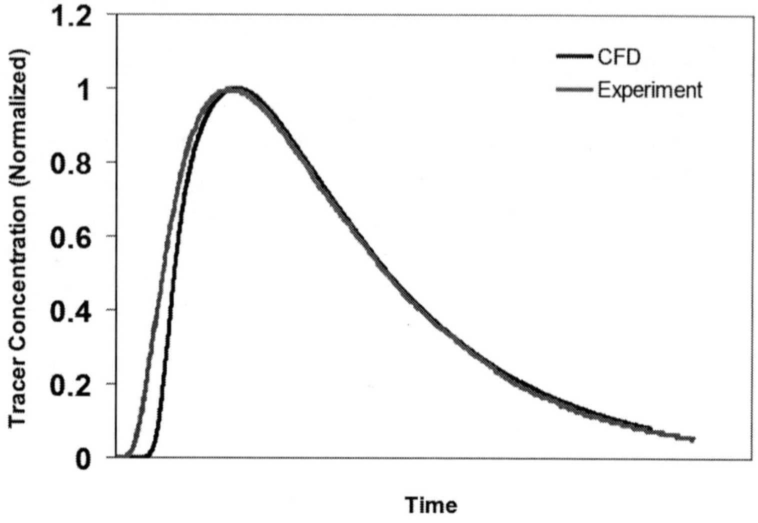

Figure 1. Comparison between the RTD curves obtained from experiments and CFD simulations for the Plexiglas model.

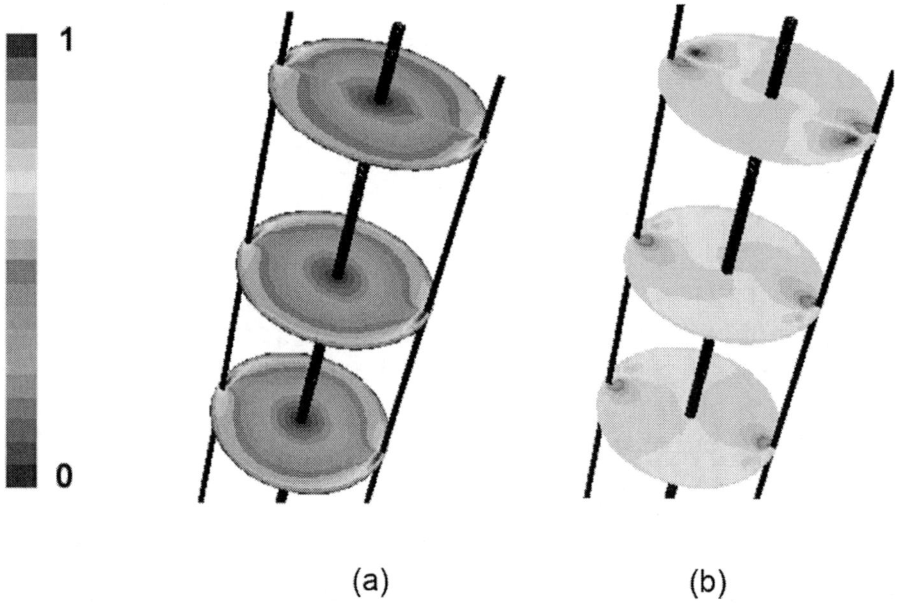

Figure 2. Velocity Profiles on three horizontal planes that cut through the Plexiglas model. The velocity is scaled as $v^* = (v - v_{min})/(v_{max} - v_{min})$. (a) Contour of the velocity magnitude. (b) Contour of the radial velocity.

The Crystallizer Model

The CFD model for the crystallizer is also validated against the RTD data as shown in Figure 3. The shape of the RTD curve predicted by the model is similar to the experimental data, but the curve is shifted slightly to the right in the simulation, which means that the tracer stays in the vessel longer in the simulation than in the experiments. This disagreement could be caused by the modifications that are made to the feed inlet and the fact that the effect of solids slip is not accounted for in this single phase model. Overall the validation is qualitatively good for this complex system.

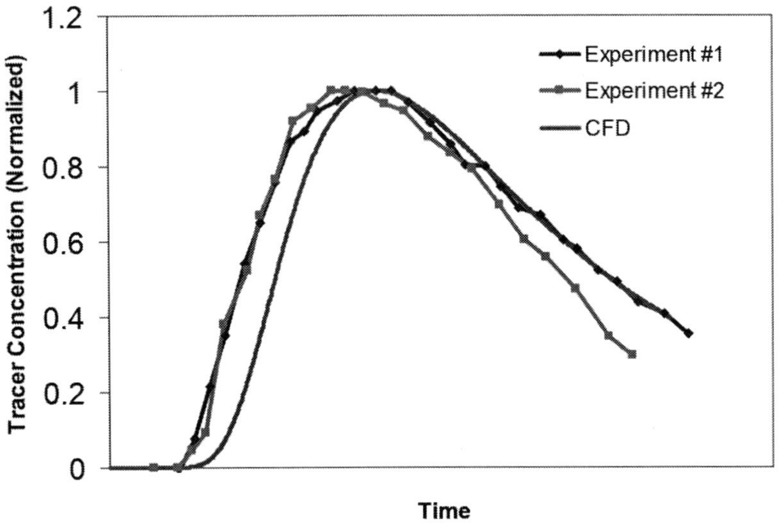

Figure 3. Comparison between the RTD curves obtained from experiments and CFD simulations for the crystallizer model.

The flow in the crystallizer is significantly different from that in the Plexiglas model due to the buoyancy effect caused by the density gradient, even though Figure 4 shows that the radial velocity profile looks similar to what is observed in the Plexiglas model. In the near wall region the density of the slurry is higher because of the presence of PX solid and lower temperature. As a result, the axial velocity is higher in this region. The animation generated during the tracer simulation shows that after the tracer is released into the vessel, part of it flows downwards without being dispersed and part of it is "drawn" to the near wall region by this higher axial velocity and then flows downwards near the wall. To show the path of the tracer flowing in the crystallizer, the contour of the tracer mass fraction at t=2000s during the tracer simulation is plotted in Figure 5. It clearly shows that the tracer moves toward the wall after it is released in the vessel (the light blue region in the top of the vessel is where the tracer is released) and it prefers to stay in the near wall region in the lower part of the crystallizer.

To quantitatively show the significance of the buoyancy effect, the ratio of the Grashof number and Reynolds number, which is the ratio of the gravity fluctuation term and the convection term in the Navier-Stokes equation, is calculated as [7],

$$\frac{Gr}{\mathrm{Re}^2} = \frac{g\beta\Delta TL}{v^2},$$

where g is the acceleration of gravity, β is the thermal expansion coefficient, ΔT is the temperature change, L is the characteristic length and v is the characteristic velocity. For the commercial crystallizers, this ratio is at the order of 1, which indicates strong buoyancy effect in the vessel. This is in agreement with the significant difference in the flow patterns observed in the crystallizer model and the Plexiglas model that was discussed earlier.

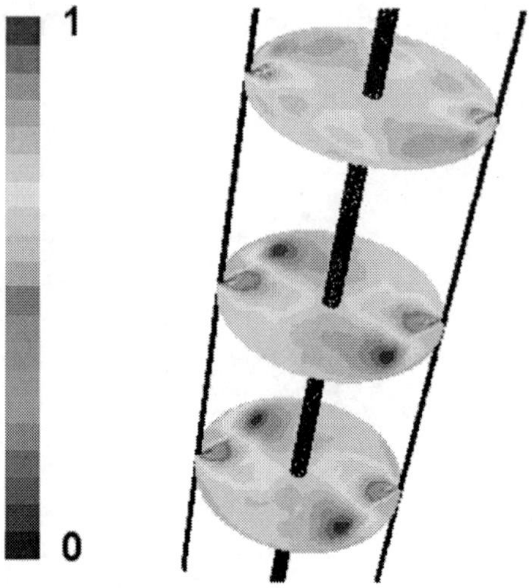

Figure 4. Contour of the radial velocity on three horizontal plans that cut through the crystallizer. Velocity is scaled as $v^*=(v-v_{min})/(v_{max}-v_{min})$.

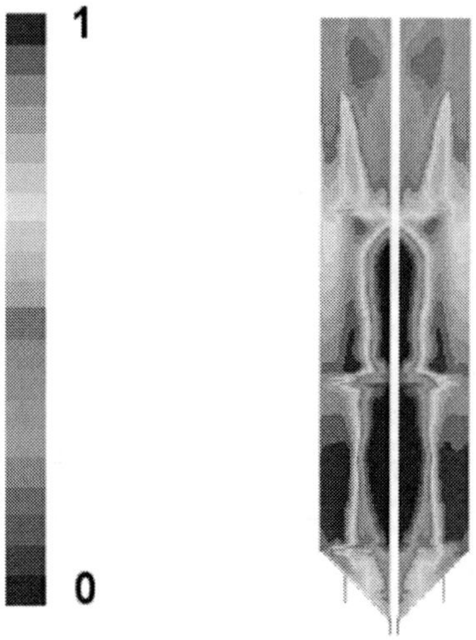

Figure 5. The contour of the tracer mass fraction at t=2000s during the tracer simulation. The concentration is scaled as $C^*=(C-C_{min})/(C_{max}-C_{min})$.

Figure 6 (a) and (b) are the contours of temperature and solid mass fraction on the vertical plane that cut through the crystallizer. Consistent with the tracer path shown in Figure 5, temperature is warmer and solid mass fraction is lower in the top center region. In the lower part of the vessel, although Figure 5 shows that the fluid in the center region is almost stagnant and the fluid flows downwards only in the near wall region, the temperature profile in the lower part of the crystallizer looks similar to the plug flow behavior. This is because the temperature of the stagnant fluid will approach that of the fluid flowing by in a steady state operation.

Therefore one cannot conclude that the flow in the lower part of the crystallizer is plug flow just based on the temperature profile.

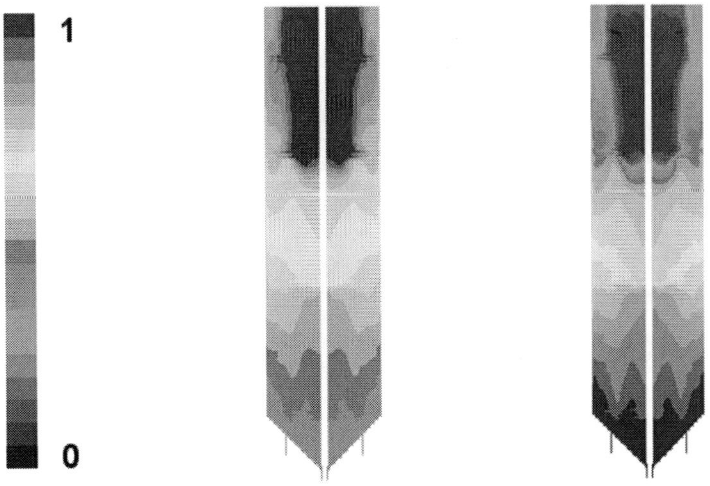

Figure 6. Contours of (a) temperature and (b) solid mass fraction on the vertical plane that cut through the crystallizer. Both variables are scaled.

Conclusions

Based on the validation of the two models for the pilot plant Plexiglas model and the commercial PX crystallizer, we can see that CFD models can predict the fluid dynamics in these vessels very well. CFD proves to be a powerful tool in studying heat transfer and fluid dynamics in these applications. In the Plexiglas model, radial velocity caused by the movement of the scraper is shown to be very low. In the PX crystallizer, the buoyancy effect completely changes the flow pattern – axial velocity is higher in the near wall region due to both the higher density caused by the presence of PX solid and lower temperature. There is a warmer region near the centre in the top half of the crystallizer, while the temperature in the lower half of the crystallizer is more uniform. However the flow in the lower half is not plug flow due to the existence of the stagnant zones near the centre.

References

1. J. F. Maingonnat. A Study of The Thermal Performance of A Scraped-surface Heat Exchanger. Part II. The Effect of The Axial Diffusion of Heat.
2. Vaessen, R. J. C., Seckler, M. M. and Witkamp, G. J. Heat Transfer in Scraped Eutectic Crystallizers. International Journal of Heat and Mass Transfer 47, 2004, pp717 – 728.
3. Qin, F. G. F, Chen, X. D. and Russel, A. B. Heat Transfer at the Subcooled-Scraped Surface with/without Phase Change. AIChE Journal, 49(8), 2003, pp1947 – 1955.
4. Zheng, L. Y., Russel, A. B. and Wilson, D. I. Influence of Crystallisation on heat transfer through a scraped surface. 6th World Congress of Chemical Engineering, Melbourn, Australia, 23 – 27 September, 2001

5. Baccar, M. and Abid, M. S. Numerical Analysis of Three-Dimensional Flow and Thermal Behaviour in a Scraped-surface Heat Exchanger. Rev. Gen. Thermodyn., 36, 782(1997)
6. Ronald de Goede. Crystallization of ParaXylene with Scraped Surface Heat Exchangers. PhD thesis. 1998.
7. FLUENT 6.3 Documentation.

143e High Temperature Co-Electrolysis of Steam and Carbon Dioxide to Produce Synthesis Gas for Fischer Tropsch Synthetic Fuels

Lyman Frost[1], Joseph Hartvigsen[2], and S. Elangovan[2]. (1) Energy and Electrochemistry, Ceramatec, Inc, 2425 South 900 West, Salt Lake City, UT 84119, (2) Electrochemistry, Ceramatec, Inc, 2425 South 900 West, Salt Lake City, UT 84119

High Temperature Co-electrolysis of Steam and Carbon Dioxide to Produce Synthesis Gas for Fischer Tropsch Synthetic Fuels

Ceramatec, Inc is currently developing technology for the high temperature co-electrolysis of steam and carbon dioxide to produce synthesis gas. This technology utilizes solid oxide fuel cell materials to produce a clean stream of carbon monoxide and hydrogen. By varying the concentrations of carbon dioxide and steam, the yielded ratio of hydrogen to carbon monoxide can be modified dependent on the type of synthetic fuel desired for production.

High temperature electrolysis is substantially more efficient than existing electrolysis methods since part of the energy required to break the molecular bonds is furnished by heat. Existing proton exchange membrane electrolysis technologies operate below 100O C but high temperature electrolysis operates in the 800O C range. This higher temperature improves the efficiency significantly. The use of the materials utilized for solid oxide fuel cells matches the desired temperature range. Issues with the materials are raised, however, since the operating conditions of the electrolysis cells vary from those of solid oxide fuel cells. While some environmental conditions are beneficial, others create potential problems that Ceramatec is addressing with its research.

Ceramatec has been working on application of this high temperature technology for production of hydrogen in association with nuclear power with the Idaho National Laboratory. The Office of Naval Research has directly funded Ceramatec to investigate the production of synthetic fuels through the co-electrolysis of carbon dioxide and steam.

This paper will present results of laboratory scale high temperature co-electrolysis. The paper will present the yields of hydrogen and carbon monoxide obtained from varying input mixtures of carbon dioxide and steam. The paper will also present the results of utilizing the resultant synthesis gas to produce synthetic methane and Fischer Tropsch liquid fuels. Issues related to long term lifetime testing, commercial scale-up, and cost will be discussed.

95e Propylene or B T X - Enhanced Production with New Technology Integration

Joseph Gentry, Technology, GTC Technology, 1001 S. Dairy Ashford Road, Suite 500, Houston, TX 77077

With the high demands of both propylene and aromatics, incremental investment within the existing petrochemical complexes could play a significant role to meet the demands and improve process economics. In this presentation, GTC will introduce aromatization, olefin catalytic cracking, and other technologies to illustrate the potential process integration and opportunities for commercial practice and R&D improvement.

143c Intermittent Feeding of Synthesis Gas into Series FT Reactors

Jim Bucher and Johannes Schwank. Chemical Engineering, University of Michigan, 2300 Hayward, 3085 H H Dow, Ann Arbor, MI 48109

The Fischer-Tropsch synthesis (FTS) produces a wide range of hydrocarbons from synthesis gas. The ability of the reaction to produce transportation fuels from coal or biomass could be vitally important in the near future due to the dwindling supply of high quality crude oil. For this reason, the current study endeavors to increase the selectivity of FTS in the C8 – C12 range.

Two series FTS reactors, both containing silica supported iron catalyst, were employed to perform this study. Synthesis gas was fed into the first reactor; additional synthesis gas was fed into the second reactor along with the products from the first. Several trials were performed in which the relative amounts of carbon monoxide and hydrogen in the secondary synthesis gas stream were varied while the CO:H2 ratio of the primary synthesis gas stream was held constant.

This presentation will discuss the results of these experiments with regards to selectivity of carbon dioxide, methane, C2 – C3 compounds, C4 – C6 compounds, C7 – C10 compounds, and C11 – C15 compounds. The affect of conversion on selectivity will also be discussed with regards to previously performed work.

Coal-derived Syngas via the KBR Transport Gasifier (TRIG™)

Siva Ariyapadi
Principal Technical Professional, Coal Monetization, KBR Houston, TX

Nicola Salazar
Director, Coal Monetization, KBR, Houston, TX

Richard Strait
Director, Technology – Asia Pacific, KBR, Singapore

Prepared for Presentation at the 2008 AIChE Spring Meeting, New Orleans, LA - April 6-10, 2008
Session #95 - Petrochemicals and Derivatives I (16B00)

KBR

Outline

- Development & Status of Transport Gasifier (TRIG™)
- Characteristics of Syngas produced
- Need for CO Shift
- Description of Process Configuration
- Configuration Benefits
- Summary

KBR

Development of TRIG™

The Transport Gasifier

Leverages KBR History of FCC Technology Expertise

- Late 30's
- 1942 — First Commercial FCC Unit for Exxon
- 1951 — Orthoflow™ A Design
- Early 1960's — Orthoflow™ C Design
- Resid FCCU
- 1976 — Orthoflow™ F Design
- 1980's — Design Based on FCC Technology
- 1990 — Pilot Plant Tech Center
- 1996 — Wilsonville, AL 8,000+ hrs of test runs
- 2005 - — Commercial Gasifier Development for IGCC Applications

KBR

Status of TRIG™

- In operation since mid-1990s (PSDF, Wilsonville, AL)
- Engineering-scale, 50 TPD demonstration Unit
- Successfully tested a wide range of Low-rank coals
- Operated in both air- and oxygen-blown modes
- Proven Technology and Operation
- 600 MW_e two-gasifier lignite-based IGCC project in detailed design phase for Mississippi Power

KBR

Transport Gasifier (TRIG™)

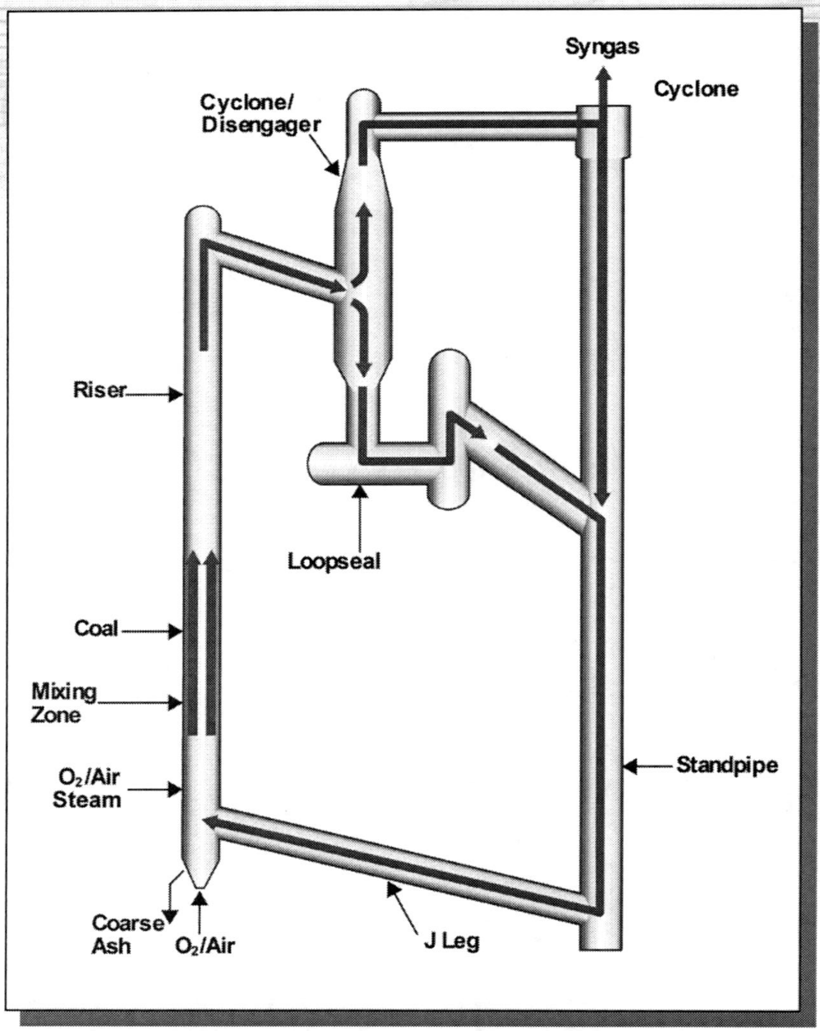

Derived from KBR's FCC Technology

Refractory Lined

Simple Design, No Internals

KBR

Particulate Control Device (PCD)

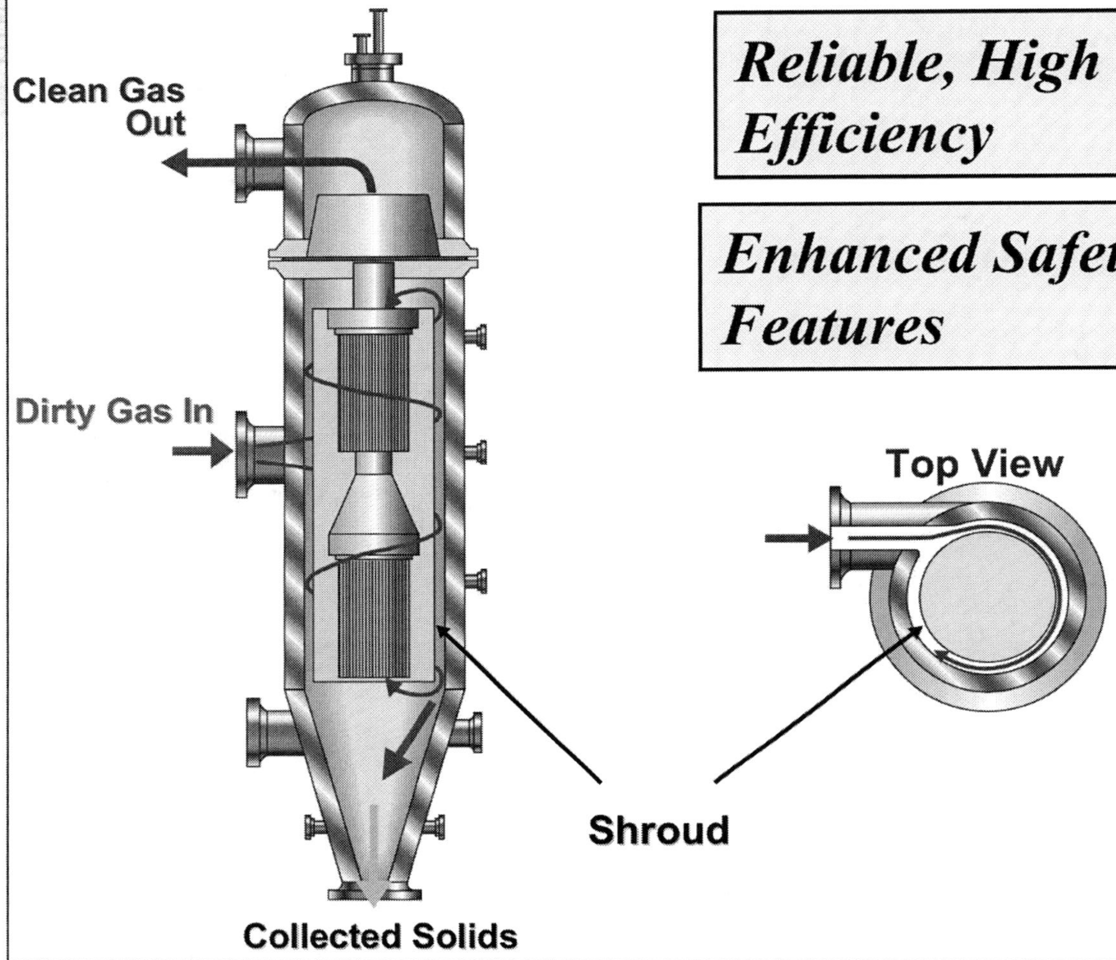

Reliable, High Efficiency

Enhanced Safety Features

KBR

Typical Characteristics of TRIG Syngas (PCD Exit)

- Steam-to-gas ratio
  - Oxygen-blown: ~ 0.2
  - Air-blown: 0.05 – 0.1
- H_2:CO ratio
  - Oxygen-blown: ~ 0.7 – 0.8
  - Air-blown: 0.5 – 0.6
- Particulate-free
- Temperature ~ 600-700°F
- Pressure ~ 500 psig

KBR

Need for CO Shift

- Target H_2:CO ratio for chemicals production:1.5:1 to 3:1
- H_2:CO ratio for H_2 production >10
- Need shifting even for power generation/CCS due to expected limits on CO_2 emissions/MW-hr
- KBR has extensive Syngas Treatment/ Shift experience from our Ammonia, Hydrogen, and Chemicals Projects

KBR

Simplified TRIG™ Process Configuration

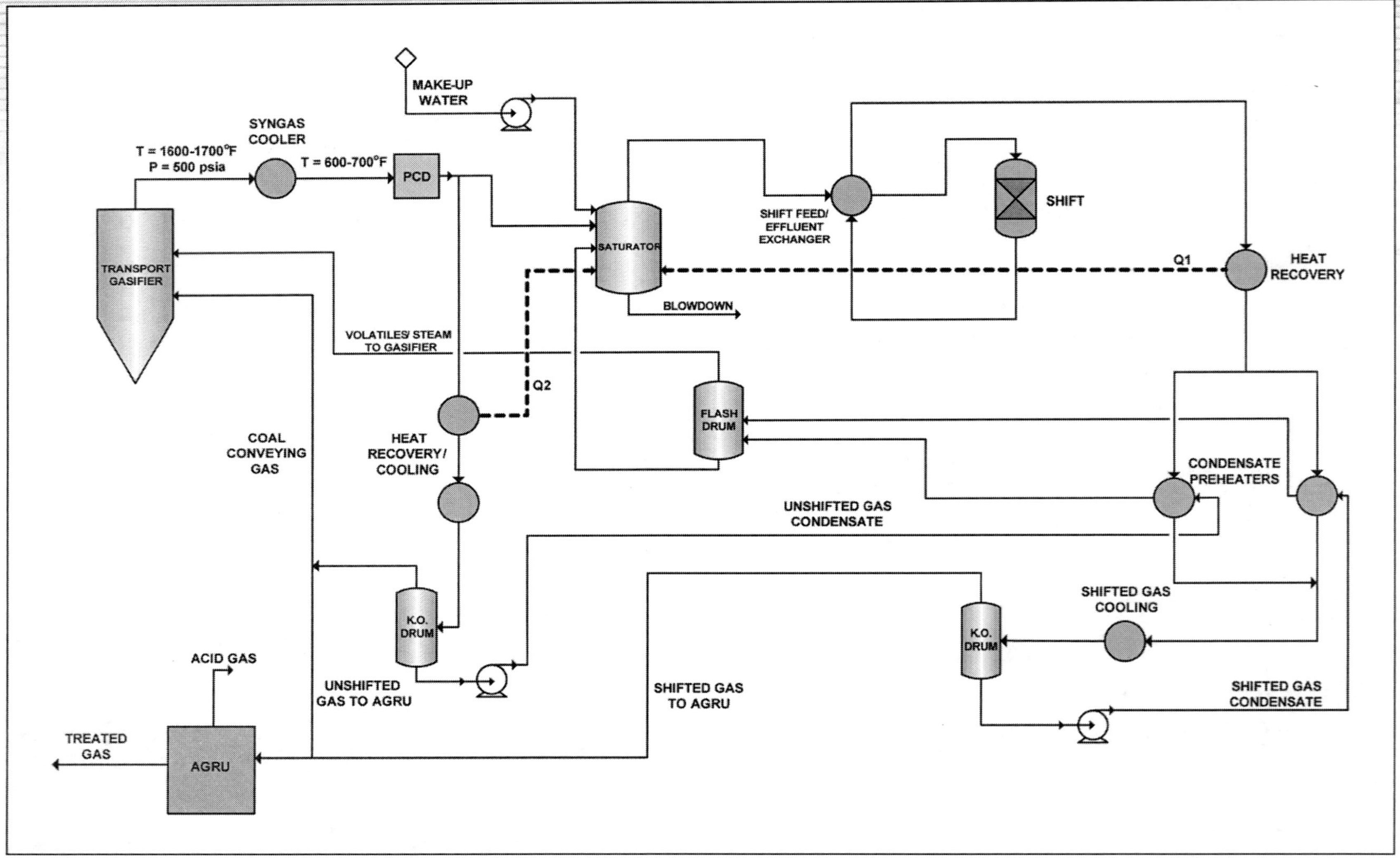

KBR

Key Benefits of Configuration

- Low specific-O_2 consumption. High cold gas efficiencies.
- No live steam usage in Shift (saturator) section
- High-grade heat primarily used for HP steam generation
- Greatly reduces water make-up and blow-down
- Volatiles recycled to gasifier and destroyed
- Enhanced Flexibility. Offers a range of target H_2:CO ratios
- Maximum recovery of both high- and low-grade heat

KBR

Summary

- TRIG™ Process Configuration is proven – builds on KBR's FCC, Ammonia & Syngas Expertise
- CO shift needed for Chemicals & Fuels, and future IGCC applications – TRIG is flexible in H_2:CO ratios
- Syngas generated is Particulate – free
- For most applications, need to saturate and shift only a portion of syngas
- Process scheme is energy-efficient, maximizes recovery of water and heat, and recycles volatiles to destruction

KBR

95b Study on the Kinetics of Ca-Based CO2 Sorbent of Carbonation and Calcination for Adsorption Enhanced Reforming Process

Su F. Wu, Qi Shi, and Qing H. Li. Department of Chemical Engineering, Zhejiang University, Zheda road No. 38, Hangzhou, China

Abstract Ca-based CO2 adsorbent may have different properties according to the kinds of application in fuel gas release, adsorption enhanced reforming (AER), energy storage and chemical pump system. Ca-based adsorbent applied in AER process needs a rapid adsorption rate and decomposition rate at lower temperature to ensure a synergic action with the methane steam reforming process. The author has systematically studied and find that nano scale Ca-based CO2 sorbent is desirable with the advantages of rapid CO2 reaction rate and high adsorption capacity, lower decomposition temperature and re-useable.

A nano scale CaCO3(70 nm) was used as precursor for CO2 adsorbent. The properties of calcination followed by carbonation was studied by TGA In order to pursue the reaction condition of as a criterion of reaction section and diffusion section, The data simulated with the Avrami-Erofeeve kinetic equation•F obtained a good results at the temperature of 500-650°C, CO2 partial pressure of 0.33atm. The simulation results showed that a relative deviation of CO2 adsorption ratio of the reaction section and diffusion section were lower than 3.5% and 0.86% respectively. The activity energy of reaction section of nano scale CaO is 11.9 kJ/mol. The activity energy of diffusion section of nano scale CaO is 106.9 kJ/mol.

Calcination kinetics of CaCO3 using equation of is in coincidence with the experimental data under the temperature of 650-800°C. At the CO2 partical pressure of 0.20 atm and 0.40 atm ,the relative deviation is lower than 7.82% and 5.01% , and the active energy of calcination reaction is 34.72kJ/mol and 113.2kJ/mol respectively.

In a summary, nano Ca-based CO2 sorbent has a lower activity energy for carbonation and followed by calcination. This advantage of nano Ca-based CO2 adsorbent is in consistence with the higher adsorption ratio and reaction rate than the micrometer scale Ca-based adsorbent. The lower decomposition reaction condition under low CO2 partial pressure and easy to be regenerated of nano scale Ca-based CO2 adsorbent are unparalleled. Also the nano Ca-based CO2 sorbent used in the AER process for producing hydrogen has been studied. Good results showed it has an attractive potential to be used in the AER process for producing hydrogen.

Keywords: hydrogen, CO2 sorbent, nano CaCO3, kinetics, adsorption, decomposition

THE EFFECT OF OXYGEN AVAILABILITY ON CYCLOHEXANE OXIDATION

R. Jevtic, P.A. Ramachandran, and M. P. Dudukovic

Department of Energy, Environmental and Chemical Engineering, Washington University in St. Louis, Chemical Reaction Engineering Laboratory
One Brooking Drive, St. Louis, 63130 USA

Introduction

Approximately 10^6 tones per year of cyclohexanol and cyclohexanone are produced worldwide by cyclohexane oxidation process[1]. This industrially important process that ultimately leads to production of Nylon 6 and Nylon 6,6 is very inefficient with cyclohexane conversion being 4-8% and the selectivity of 80% for cyclohexanone and cyclohexanol[2]. Despite all the research efforts, the technology of cyclohexane oxidation process has not been improved significantly and the basic process that was pioneered by DuPont in 1940s[3] is still largely in use.

Greene et al[4] reported the results of cyclohexane oxidation with pure oxygen performed in the Liquid Oxidation Reactor (LOR). The residence time was reduced from 36 to 8 min, temperature of operation was slightly reduced, and selectivity and productivity are increased while keeping the same cyclohexane conversion of 4%. Since the LOR results were compared with the data from conventional process, it is not clear whether the improvements were due to the increased oxygen availability or to the different transport effects in the reactor itself. To assess this, one needs a complete set of data on conversion and selectivity as a function of time.

Experimental Section

Experiments were performed in a 25ml Parr® autoclave equipped with Parr® 4843 controller. The reactor was operated in the batch and semibatch mode. In first case, the reactor was charged with 7.5 ml of cyclohexane then pressurized with nitrogen, heated to the reaction temperature of 130^0C, and finally the oxygen would be added to make up the total pressure of 15 bar. When the reactor was operated in the semibatch mode, the initial procedure was the same. But in this case, reactor was equipped with the solenoid vale, additional pressure transducer and the process control unit. When the total pressure in the reactor would drop below the set point, due to oxygen consumption, the valve would open and oxygen would be supplied. The idea behind this is to keep the partial pressure of oxygen in the reactor constant. The experimental runs were performed without the use of catalysts and with the use of catalyst. In the latter case, cobalt naphthanate was used.

Quantitative analysis of products is performed using Agilent 6890 Gas Chromatograph (GC) with Flame Ionization Detector (FID). Helium was used as a carrier gas. A 30m Alltech® AT-5ms column (0.25mmx0.25mm) was used for separation of oxidation products. The following measurement conditions were used: injection port temperature was held at 220^0C. Initial temperature of the column was 70^0C with immediate rise in temperature of 10^0C /min for 10 min. After that, when temperature reached 170^0C, final rise was 50^0C /min up to 230^0C where it was held for 2 min. The MSD

Productivity ChemStation Software was used to record and integrate the output from the FID.

Modeling

Experimental studies are time consuming and, thus, they are usually used for a limited number of reaction conditions. On the other hand, models that capture well the reaction in a given experimental set up can be then used to predict the outcome in any other condition of interest. To set up appropriate reactor model, we started with the simple one. Assuming constants temperature, fully back mixed gas and liquid phase, validity of ideal gas law for the gas phase and occurrence of the reactions in the liquid phase only, species balance in the gas phase for the batch operation is given as

$$\frac{V_G}{RT}\frac{dp_i}{dt} = -k_L a\left(\frac{p_i}{H_i} - c_i\right)V_L \tag{1}$$

Species balance in the liquid phase is:

$$\frac{dc_i}{dt} = k_L a\left(\frac{p_i}{H_i} - c_i\right) + \sum_{k=1}^{NR} \nu_{ki} r_k \tag{2}$$

In the equations above, V_G is the volume of the gas phase in the reactor, p_i is partial pressure of the component i, R is the gas constant, T is reaction temperature, t is time, k_L is the mass transfer coefficient from the liquid side obtained from literature[5], a is the ratio of surface area to volume of the liquid, c_i is the concentration of the component i in the liquid phase, H_i is Henry's constant ($H_i = \frac{p_i^{in}}{c_i^{in}}$) for component i determined using Aspen Plus, V_L is the volume of the liquid phase in the reactor, ν_{ki} is the stoichiometric coefficient of the component i in the reaction k, and r_k is the reaction rate of the reaction k.

For semi-batch mode of operation, oxygen is being supplied as it gets consumed in the reactor. Since cyclohexane is very volatile, and its partial pressure depends on the concentration in the liquid phase, for higher conversions of cyclohexane it is not correct to assume that gas phase is in the pseudo-steady state. So, the species balances for gas and liquid phase remain the same as for the batch reactor with difference in the balance of oxygen in the gas phase. Oxygen balance in the gas phase is given as:

$$\frac{V_G}{RT}\frac{dp_{O2}}{dt} = -k_L a\left(\frac{p_{O2}}{H_{O2}} - c_{O2}\right)V_L + P_T - \sum_{i=1}^{NC} p_i \tag{3}$$

The additional terms account for oxygen that is being provided to the reactor when total pressure in the system drops below set point (P_T).

For simplicity, the kinetic scheme used in the models is the one proposed by Spielman (1964)[6] and Alagy et al. (1964)[7] that consists of lumped reactions without free radicals:

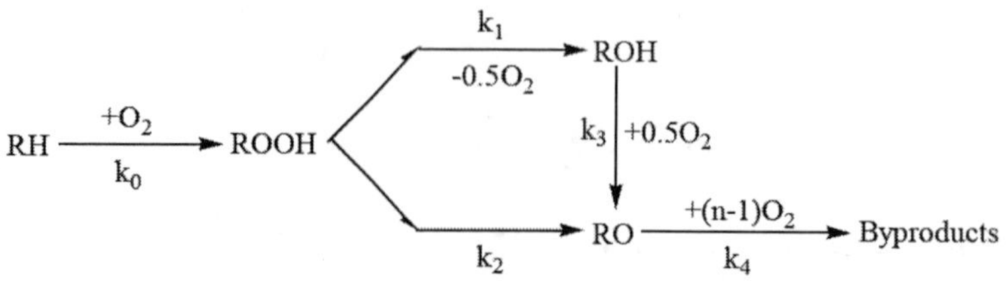

The kinetic constants for this kinetic scheme are obtained from the literature[8] (Table 1).

Table 1. Kinetic constants from the literature

k_0 ($m^3kmol^{-1}s^{-1}$)	k_1 (s^{-1})	k_2 (s^{-1})	k_3 ($m^3kmol^{-1}s^{-1}$)	k_4 ($m^3kmol^{-1}s^{-1}$)
0.0266	0.0048	0.0013	0.5	0.37

These reaction constants have been estimated from the operation of the industrial reactor at 160^{0}C. There are two issues with these kinetic constants. First, it is very likely that these constants are not intrinsic since there has not been provided a proof that the reactor operated at kinetic control regime. Second, the constants are not given in the Arrhenius form so they can not be applied at different temperatures. Hence, these kinetic constants are used as initial guesses for our system.
The model equation for the gas and liquid phase are solved simultaneously with all parameters. The stiff ODE solver from Netlib library (LSODE) is used. To determine the kinetic constants, ODRPACK, a collection of Fortran subroutines for fitting a model to data, also available from Netlib, is used. Orthogonal distance Regression is a method for linear regression that determines the values of unknown quantities in a statistical model by minimizing the sum of the squared residuals - the orthogonal difference between the predicted and observed values.

Results and Discussion

First, cyclohexane oxidation was performed in the batch system with 20% and then with 50% oxygen on the gas phase. Summary of the experimental results on cyclohexane oxidation in the batch system with different concentrations of oxygen in the gas phase is provided in Table 2. Maximum conversion of cyclohexane and maximum yield of desired products are higher when more oxygen is in the gas phase. However, selectivity is lower. This was expected since it is known that selectivity decreases in cyclohexane oxidation with an increase in cyclohexane conversion

Table 2. Summary of the experimental results on cyclohexane oxidation performed in the batch mode with 20% and 50% of oxygen in the gas phase

Oxygen in the gas phase (%)	Maximum conversion (%)	Maximum yield (%)	Selectivity (%)
20	2.0	1.8	90.2
50	4.5	3.4	76.7

What remains to be confirmed is whether selectivity and productivity changes with increased availability of oxygen at the same level of cyclohexane conversion. The best way to assess this is to keep partial pressure of oxygen constant since the effect of oxygen availability is the parameter that is being examined.

Experimental results for cyclohexane oxidation with different oxygen content in the gas phase (20% and 50%) in semi-batch system are presented in Figure 1 and Figure 2. The same reaction conditions were employed as with the batch study: 130^{0}C, 15 atm, no catalyst. Again, higher yields and higher cyclohexane conversion are achieved with more oxygen in the gas phase, but, also, selectivity is lower. Overall, it is evident that reaction proceeds faster with 50% oxygen. After 30 min, cyclohexane conversion is around 7% if 50% oxygen is used as opposed to around 3% when 20% oxygen is in the gas phase. From this, one can conclude that oxygen concentration does have an effect on the rates of reaction. Still, it is unclear by how much the productivity is increased with increased oxygen availability since it is very difficult to stop experiments at the same level of cyclohexane conversion. That is done easily with the model providing that the model is successful in capturing the reaction trends and outcome in the experimental reactor.

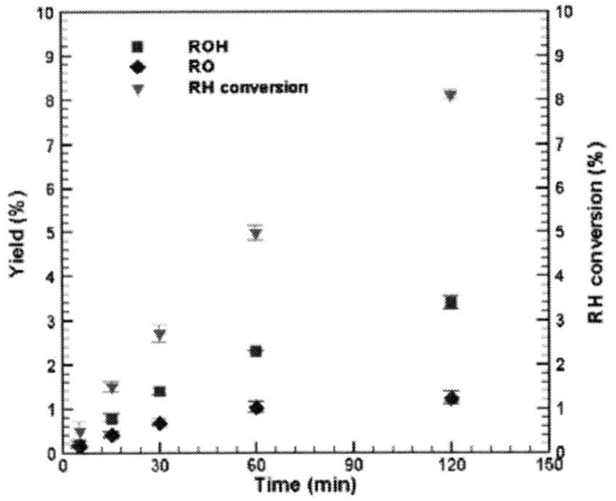

Figure 1. Experimental results of the yields of the main products and the conversion of cyclohexane in cyclohexane oxidation performed in the semi-batch reactor at 130^{0}C and 15 atm with **20% oxygen** in the gas phase

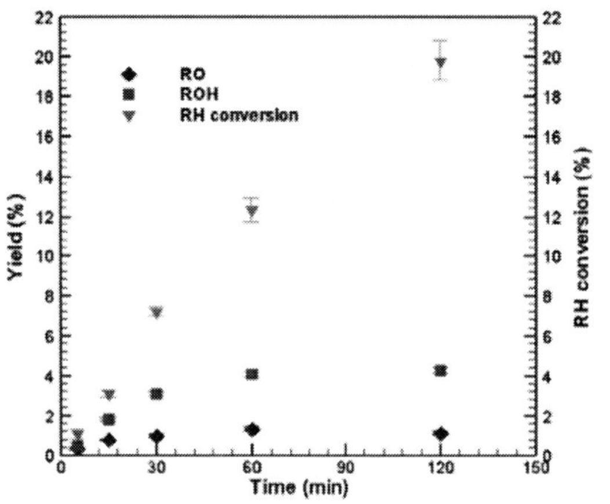

Figure 2. Experimental results of the yields of the main products and the conversion of cyclohexane in cyclohexane oxidation performed in the semi-batch reactor at 130^{0}C and 15 atm with **50% oxygen** in the gas phase

With the use of ODRPack, the kinetic constants for the system under investigation were estimated and used thereafter in the study. Figure 2 shows the experimental data for the batch operation of cyclohexane oxidation that are fitted with the model explained in the modeling section. From the model, new kinetic constants are obtained for the kinetic scheme used in this study.

In addition, these kinetic constants can be used to check the validity of semibatch model by comparing the modeling and experimental results. This is shown in Figure 3 and Figure 4. Since the model based on the previously obtained kinetic constants agrees with the experimental results, this model can also be used to estimate the progress of the reaction at desired points. If cyclohexane oxidation is to be performed at various oxygen concentrations in the gas phase and the reaction stopped at 4% cyclohexane conversion, the selectivity for cyclohexanol and cyclohexanone and the residence

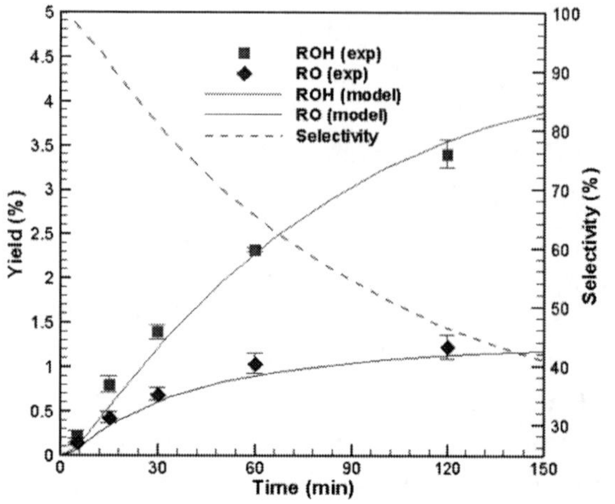

Figure 4. Comparison of experimental and modeling results of the yields of the main products in cyclohexane oxidation performed in the semi-batch reactor at 1300C and 15 atm with **50% oxygen** in the gas phase.

licted by the model, are shown in Table 2. From there, ailability in cyclohexane oxidation, the residence time 5 times shorter when oxygen is increased from 20% to creased as well. Also, the selectivity for two desired increased with an increase in oxygen availability.

at 4% cyclohexane conversion and time needed to achieve that conversion at different oxygen percentage in the gas phase predicted by the model

% oxygen in the gas phase	Time (min) to achieve 4% cyclohexane conversion	Selectivity (%) for cyclohexanol and cyclohexanone
20	43.6	73.4
50	15.6	78.9
75	9.6	82.3
100	7.7	84.9

Conclusions

To examine the effect of oxygen availability on cyclohexane oxidation, the reaction is performed in the batch and the semi-batch system where oxygen is supplied as it gets consumed in the reactor. The oxygen content in the gas phase was 20% and 50%. The isothermal models for the batch and semi-batch operated systems are solved and by fitting to the experimental data, kinetic constants at different temperatures for cyclohexane oxidation are obtained. It can be concluded from the experimental and the modeling results that increased oxygen availability improves the productivity of cyclohexane oxidation at the fixed cyclohexane conversion (4%) - residence time to achieve such conversion declines with the increase of oxygen concentration. Also, the selectivity for two desired products, cyclohexanol and cyclohexanone, improves at the same time. These findings agree with the findings of Greene and his collaborators.

References

1) Schuchardt, U.; Calvarlho, W. A.; Spinace, E. V., Why is Interesting to Study Cyclohexane Oxidation? *Synlett* **1993,** 10, 713-718.
2) Suresh, A. K.; Sharma, M. M.; Sridhar, T., Engineering aspects of industrial liquid-phase air oxidation of hydrocarbons. *Ind. Eng. Chem. Res* **2000,** 39, 3958-3997.
3) Loder, D. J. Production of Cyclic Alcohols and ketones. 2223494, 1940.
4) Greene, M. I.; Sumner, C.; Gartside, R. J. Cyclohexane oxidation. 5,780,683, 1998.
5) Tekie, Z.; Li, J.; Morsi, B. I., Mass Transfer Parameters of O2 and N2 in Cyclohexane under Elevated Pressures

and Temperatures: A Statistical Approach. *Ind. Eng. Chem. Res.* **1997,** 36, (9), 3879-3888.
6) Spielman, M., Selectivity in hydrocarbon oxidation. *A.I.Ch.E. J.* **1964,** 10, (4), 496-501.
7) Alagy, J.; Defoor, F.; Franckowiak, S., Oxidation of cyclohexane to cyclohexanone. *Rev. Inst. Franc. Petrole Ann. Combust* **1964,** 19, (12), 1380-1390.
8) Pohorecki, R.; Baldyga, J.; Moniuk, W.; Krzysztoforski, A.; Wojcik, Z., Liquid-phase oxidation of cyclohexane - modeling and industrial scale process simulation. *Chem. Eng. Sci.* **1992,** 47, (9-11), 2559-64.

95h Optimization and Control of An Industrial Packed-Bed Reactor for the Production of Ethylene Oxide

Omar Galán, Chemical Engineering, Louisiana State University, 110 Chemical Engineering, South Stadium Road, Baton Rouge, LA 70803, V. G. Gomes, Chemical Engineering, Sydney University, CNR shepherd and Lander St., Sydney, Australia, and Jose Romagnoli, Cain Department of Chemical Engineering, Louisiana State University, 110 Chemical Engineering, South Stadium Road, Baton Rouge, LA 70803.

This work tackles the control problem of an existing industrial multi-tubular packed-bed reactor for the production of ethylene oxide. The production of ethylene oxide is based on the partial oxidation the ethylene with oxygen. Complete combustion of ethylene, as well as further oxidation of ethylene oxide results in the formation of carbon dioxide and water (no desirable products). This specific control problem can be re-formulated as the steady production of the ethylene oxide without compromising personnel safety and equipment integrity. However, temperature excursions at dangerous levels have been experienced due to variations on the inlet oxygen concentration in the fresh feed to the reactor and other no measurable disturbances. The stabilization of the process is achieved by manipulation of the inlet coolant and feed temperatures to the reactor and the addition of 1, 2-dichloroethene that acts as reaction moderator. Currently, the physical sensors for gas pressure and composition are located in the inlet and outlet streams of the reactor, respectively. The temperature, pressure and oxygen concentration measurements may be regarded as real-time, suitable for control purposes. However, the delayed response from the concentration analyser for ethylene oxide and other reaction by-products is suitable only for process monitoring. Since, the chemical reactions involved in the synthesis of ethylene oxide are highly exothermic, it makes the internal temperature control of this reaction unit a challenging task due to dangerous transient responses of the catalyst bed core temperature. The lack of real-time measurements for composition and temperature along the reactor handicaps the effort to implement control strategies. In practice, a model-based control strategy requires real-time measurements of the temperature and composition profile along the reactor which are not available in this reaction unit. In order to exploit the fact that some measurements are available, the problem associated with the need of real-time measurements along the reactor can be overcome with the design of an observer. This method combines a priori knowledge about the system (plant data) with a mathematical model to provide a real-time estimation of the temperature and composition profiles along the reaction vessel. The observer synthesis derives from traditional Extended Kalman Filter and more recent approaches based-on high-gain techniques. The observer or soft-sensor is capable of estimating the ethylene oxide concentration in the reactor outlet stream and the catalyst bed temperature along the reactor from the real-time measurements. The observer is also corrected when the composition analyser produces an output. The observer-based control is robust and it presents a good closed-performance despite modelling mismatch and sustained disturbances.

Microalgae Cultivation in Bioreactors for CO_2 Mitigation from Power Plant Flue Gas and Fuel Production by Supercritical CO_2 Extraction

Maoqi Feng, Division of Chemistry and Chemical Engineering, Southwest Research Institute, San Antonio, TX 78238, email: mfeng@swri.org.

Abstract: Mitigation of CO_2 from the flue gas of coal-fired power plant is gaining more and more attention. Microalgae provides a pathway for such application because microalgae consumes CO_2 gas as a carbon source for their growth by photochemical synthesis. After the CO_2 from the flue gas is separated, a part of it can be used as a carbon source for microalgae cultivation. Botryococcus braunii (*B. braunii*) is selected as the alga because of its high hydrocarbon content and high CO_2 concentration tolerance. Cultivation of *B. braunii* in bioreactors utilizes CO_2 separated from the flue gas and solar energy as the energy source at daytime and artificial light at nighttime. Optimizing the culture conditions will increase CO_2 fixation rates. Hydrocarbon species extracted from *B. braunii* are summarized and tabulated. There are several different ways for fuel production from microalgae, and two methods were discussed: 1) hydrocarbon extraction with organic solvents or supercritical CO_2, and subsequent pyrolysis or catalytic cracking of the extracts is an efficient way to produce high quality transportation fuel; 2) direct thermochemical liquefaction, which operates at high temperature (300 °C –350 °C) and under catalyst, e.g., sodium carbonate, eliminates the need for initial extraction step for recovery of hydrocarbons, but the overall yield of oil is lower. The application of supercritical CO_2 as solvent for hydrocarbon extraction from microalgae appears more economic than organic solvent extraction, as the captured CO_2 from power plant flue gas can be condensed and then used for this application, and the process is solvent-free, eliminating organic solvent, thus avoiding the production of waste solvents. Supercritical CO_2 extraction in nanofiltration membrane reactor might improve the extraction efficiency.

1. Introduction

Electrical power plants are responsible for more than one-third of the US emissions, and for about 7% of the world's CO_2 emissions from energy use. Mitigation of the CO_2 levels emitted from flue gas of power plants can be done by biological CO_2 fixation and utilization using photosynthetic microorganisms, such as microalgae. Photosynthetic microorganisms utilize flue gas CO_2 as a carbon source and solar energy as the energy source to produce biomass, which can be used for fuel production. From 2000 to 2005, DOE funded a project with this concept, which is for CO_2 recovery and sequestration to reduce CO_2 emission from power plants by photosynthesis of microalgae [1].

Figure 1 shows the diagram of CO_2 sequestration and utilization from flue gas by photosynthesis of microalgae. The CO_2 from flue gas emitted by coal-burning power plants is treated as raw material in the system. The CO_2 is separated from the flue gas at first, and then it is fixed within a photobioreactor by photosynthesis using microalgae.

Sunlight can be used for photosynthesis in the photobioreactor. The algae in the photobioreactor will fix CO_2. The algae is collected and extracted with supercritical CO_2, and the hydrocarbons extracted from algae will be cracked to produce useful liquid fuel.

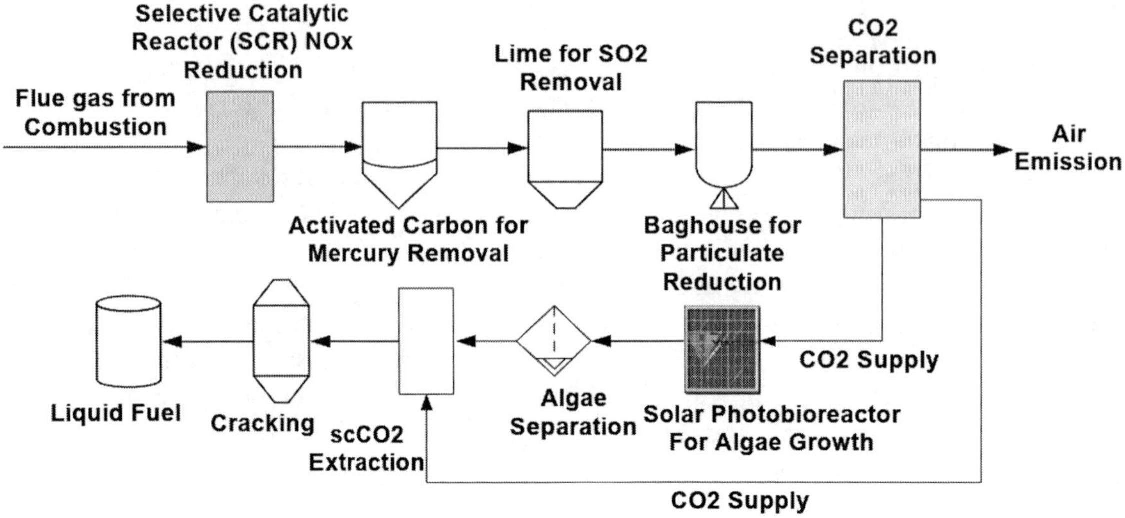

Figure 1. Process flow diagram for sequestration and utilization of CO_2 from power plant flue gas by photosynthesis of microalgae

The cost of removing CO_2 from a conventional coal-fired power plant with flue gas desulfurization was estimated to be in the range of $35 to $264 per ton of CO_2 [2]. DOE's goal is to reduce the cost of carbon sequestration to below $10/ton of avoided net cost. The valuable fuel produced from the harvest algae will greatly reduce the cost of CO_2 mitigation for power plant. Though CO_2 cost plays an important role in the economics of microalgae cultivation [3], CO_2 source from flue gas is very cost-effective. Table 1 lists the facility size needed for algae productivity for a 500 MW coal-fired power plant.

Table 1. A 500 MW coal-fired power plant [3]

Parameter	Value
Coal consumption, t/d	5,556
Flue gas emission, MM SCFD/MW	2.25
Flue gas flow rate, SCFM Sm3/min	781,250 22.503
CO_2 concentration, vol.%	14.0
Total CO_2 recovered, t/yr	2,830,311
CO_2 recovery, %	88.6
Equivalent microalgae facility size for algae productivity of 45 g/m^2d, ha	13,862

2. Microalgae Selection

Extensive screening had been conducted in literature to obtain photosynthetic microalgae from nature with high capability in fixing CO_2 and/or with the ability to produce useful materials. In Reference [1], 34 different strains were analyzed for high value pigments, and the productivity parameters for 25 different algae grown were determined. Two green algal strains obtained by the screening, Chlorella sp. UK001 and Chlorococcum littorale, showed the high CO_2 fixation rates exceeding the initial target value of 1g CO_2/1/day at 24 h illumination [4]. One alga, Botryococcus braunii (B. braunii), has received extensive attention due to their unique characteristics, high CO_2 tolerance during growth, and where hydrocarbons can account for up to 75% of the dry mass, making them a potential fuel resource [5]. The yield of liquid fuel produced from B. braunii and its heating value is high compared with other algae, e.g., Dunaliella tertiolecta.

B. braunii is a unicellular photosynthetic green colonial microalga, which is widely distributed in freshwater and brackish water. It is a member of the chlorophyceae (chlorophyta). In US, the Florida Everglades region is the largest subtropical wetland for favor of B. braunii growing [6]. B. braunii can be classified into three races: A, B, and L, according to the types of hydrocarbons they produce. Race A of B. braunii produces n-alkadienes and trienes with odd-carbon numbers ranging from C25 to C31; Race B of B. braunii produces two types of triterpene hydrocarbons, known as botryococcene components (C_nH_{2n-10}, n = 30–37) and a small amount of methylated squalenes (C31–C34); Race L of B. braunii produces a single tetraterpenoid, lycopadiene (C40). Historically, Race B of B. braunii was studied more than the other two because of its geochemical significance [5]. Figure 2 shows the algal microstructure of *B. braunii* [6]. As we can see, there are many oil particles in the algal cells. Efforts are now under way to increase their CO_2 fixation rates by optimizing the culture conditions and to develop applications for the products [4].

If a 100 MW coal-fired power plant would be replaced by liquid fuel produced from B. braunii, the quantity of CO_2 mitigation could be 1.5×10^5 t/year and 8.4×10^3 ha of microalgal cultivation area could be necessary [7].

Reference [4] selected B. braunii SI-30 as the candidate for mass culture, also because of its high CO_2 fixation rate and hydrocarbon content. The CO_2 fixation rate exceeded the target value of 1 g CO_2/L/day at 24 h illumination under small-scale culture. The optimum growth temperatures for B. braunii are between 25 – 30 °C [4].

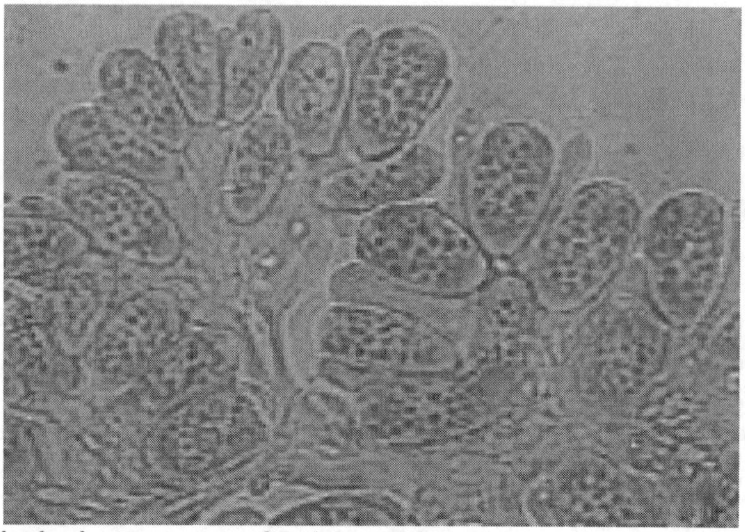

Figure 2. The algal microstructure of *B. braunii* [8]

3. Hydrocarbons in Microalgae

This paper focus on hydrocarbon production from algae, which is produced from power plant CO_2 sequestration and utilization. Therefore, it is important to know what hydrocarbons are in the selected alga, *B. braunii*.

In 1968, the hydrocarbon fraction of *B. braunii*, growing in natural environment, was isolated and examined. Surprisingly it was found that the hydrocarbon fraction comprised 76% of the dry weight of the alga. Two unsaturated isomeric hydrocarbons of formula $C_{34}H_{58}$, botryococcene and isobotryococcene, constituted the bulk of the hydrocarbons. No saturated hydrocarbons were detected [9].

In 1969, three physiological states, characterized by their hydrocarbon content were described for the green alga *B. braunii*: (1) Green active state colonies containing a complex mixture of hydrocarbons of the general formula C_nH_{2n-2} and C_nH_{2n-4}, (2) Brown resting state colonies containing a high concentration of a nearly pure (90%) hydrocarbon botryococcene, (3) Large green cells showing very little synthesis of hydrocarbons [10].

In 1983, a branched chain hydrocarbon ($C_{36}H_{62}$, darwinene) of the botryococcene type was isolated from *B. braunii*. The compound was the major hydrocarbon component from colonies collected in the Darwin River Reservoir in Australia [11].

In 1985, nine branched hydrocarbons of the botryococcene type (C_nH_{2n-10}, $30 \leq n \leq 37$) were isolated from *B. braunii*. Hydrocarbon mixtures were recovered from wild algae collected in fresh water lakes or from the same strains growing in laboratory; they were further separated by reversed-phase, and in some cases by normal phase, HPLC. From chemical investigations, GC-MS analyses, ^1H and ^{13}C NMR spectroscopy, the structures of four new botryococcenes (one $C_{33}H_{56}$, two $C_{34}H_{58}$ and one $C_{37}H_{64}$) were elucidated [12].

In 1988, a cyclic hydrocarbon of the botryococcene type, Meijicoccene ($C_{32}H_{54}$), was isolated from the colonial green alga *B. braunii* strain Berkeley mass-cultured at Meiji University of Japan. This compound was a minor hydrocarbon (10% of total hydrocarbons) and was separated from other botryococcenes by reversed-phase HPLC [13].

Table 2 lists the major characteristic hydrocarbons produced by race B of *B. braunii*, and the hydrocarbon constituents are listed in Table 3.

Table 2. The major characteristic hydrocarbons and their structures produced by race B of *B. braunii* [5]

Carbon Numbers	Molecular formula
C25	$C_{25}H_{48}$, pentacosa-1,16-diene $CH_3(CH_2)_7CH{=}CH(CH_2)_{13}CH{=}CH_2$
C27	$C_{27}H_{48}$, $CH_3(CH_2)_7CH{=}CH(CH_2)_{11}CH{=}CH\text{-}CH{=}CH\text{-}CH{=}CH_2$ $C_{27}H_{51}$, $CH_3(CH_2)_7CH{=}CH(CH_2)_{13}CH{=}CH\text{-}CH{=}CH_2$ $C_{27}H_{52}$, heptacosa-1,18-diene, $CH_3(CH_2)_7CH{=}CH(CH_2)_{15}CH{=}CH_2$
C29	$C_{29}H_{54}$, nonacosa-1,20,22-diene, $CH_3(CH_2)_5CH{=}CH\text{-}CH{=}CH(CH_2)_{17}CH{=}CH_2$ $C_{29}H_{56}$, nonacosa-1,20-diene, $CH_3(CH_2)_7CH{=}CH(CH_2)_{17}CH{=}CH_2$
C30	$C_{30}H_{50}$, $(CH_3)_2C{=}CH(CH_2)_2C(CH_3){=}CH\text{-}(CH_2)_2C(CH_3)(CH{=}CH_2){=}CHCH(CH_3)(CH_2)_2CH{=}CH(CH_3)CH_2CH_2CH{=}C(CH_3)_2$
C31	$C_{31}H_{58}$ $C_{31}H_{60}$, hentriaconta-1,22-diene, $CH_3(CH_2)_7CH{=}CH(CH_2)_{19}CH{=}CH_2$
C30	$C_{30}H_{50}$ 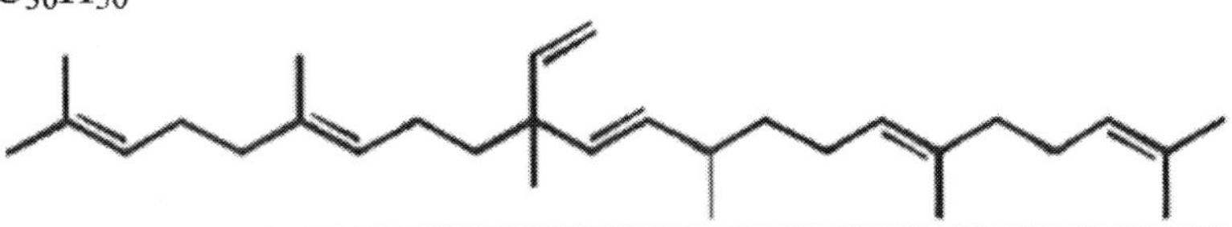
C31	$C_{31}H_{52}$ (isowolficine) and $C_{31}H_{52}$ isomer

	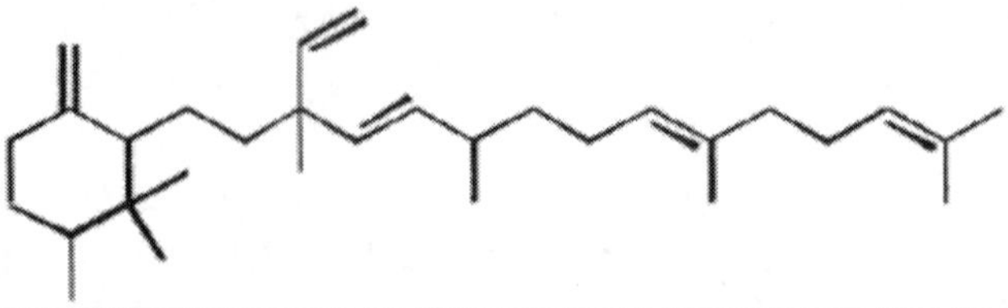
C32	$C_{32}H_{54}$ (Meijicoccene/Braunicene), $C_{32}H_{54}$ (isobraunicene) and $C_{32}H_{54}$ isomer 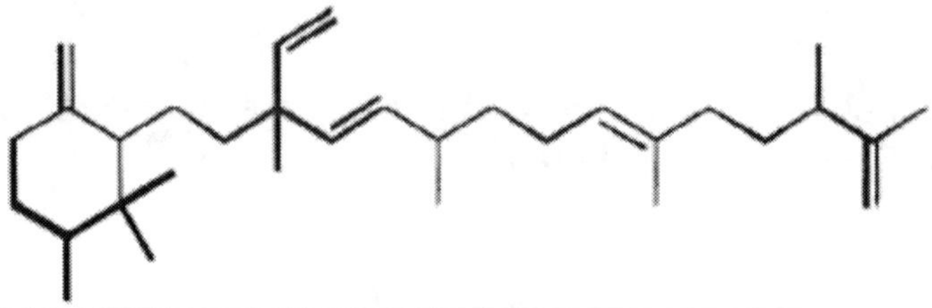
C33	$C_{33}H_{56}$ 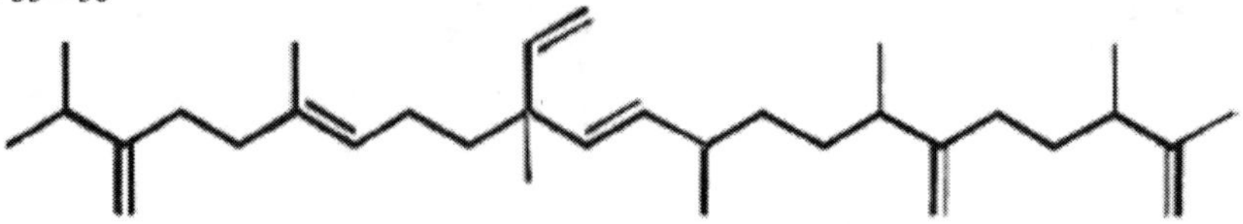
C34	$C_{34}H_{58}$ (isobotryococcene) and 4 different isomers 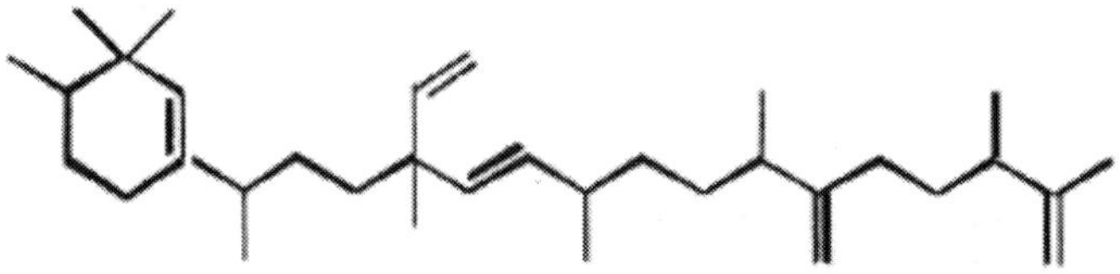
C35	$C_{35}H_{60}$, structure undecided
C36	$C_{36}H_{62}$ (Darwinene) 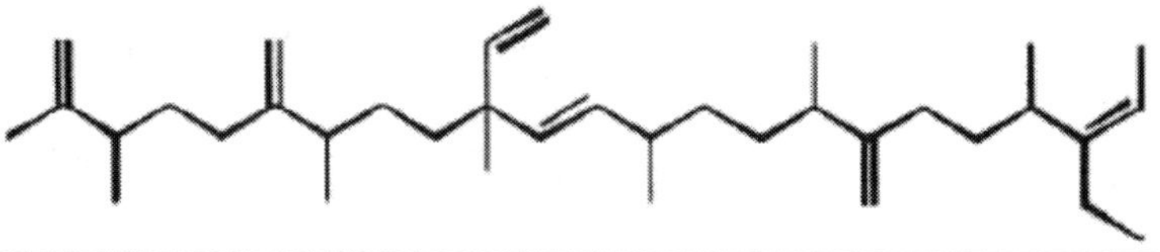
C37	$C_{37}H_{64}$ 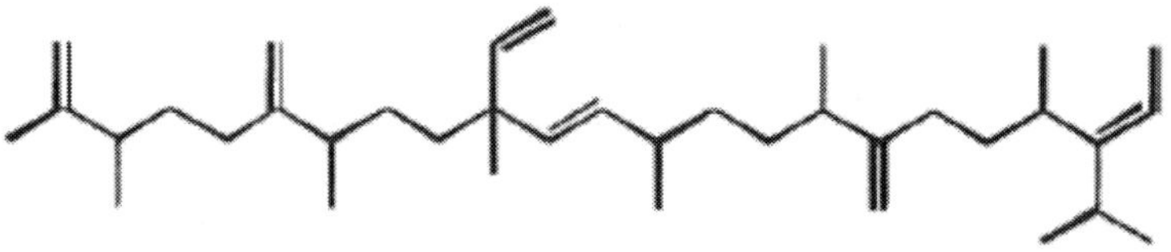

Table 3. Hydrocarbons constituents of *B. braunii* [14]

Compound	mass%
Isobotryococcene	4%
Botryococcene	9%
$C_{34}H_{58}$	11%
$C_{36}H_{62}$	34%

$C_{36}H_{62}$	4%
$C_{37}H_{64}$	20%
Other hydrocarbons	18%

4. Microalgae Growth

Microalga *B. braunii*, which produces high levels of liquid hydrocarbons, grows in fresh water or saline water. Both natural pond and photobioreactor can be used for *B. braunii* harvest. In a photobioreactor, solar light can be used for daytime culture, and artificial lights can be used for nighttime harvest. In Reference [1], pilot scale photobioreactors (0.18 m diameter, up to 2,000 liter capacity) were tested for eight microalgal strains. One of them is UTEX LB572 *B. braunii*. Different cultivation conditions: temperature, light and salinity were studied [8].

Bubble column photobioreactors were used to cultivate *B. braunii* in Reference [15]. Algal cells, which were adapted to low irradiance (3 klx) in the preculture, showed lower biomass and hydrocarbons productivity in the photobioreactor illuminated at high irradiance (10 klx) due to the effects of photoinhibition. The degree of photoinhibition was reduced by partial shading, the lighted volume ratio being varied from 25 to 100%. The algal cells adapted to high irradiance in the preculture showed high biomass productivity at 10 klx: the cell concentration reached higher than 7 kg/m^3 and the percentage of hydrocarbons was 50 wt.% based on cell dry weight. Hydrocarbon production kinetics for *B. braunii* in the linear growth phase was found to be growth-associated irrespective of the experimental conditions of illumination and preculture [15].

5. Hydrocarbon Extraction with Organic Solvent

Once the *B. braunii* biomass is recovered from the broth by filtration or centrifugation, the hydrocarbons in the *B. braunii* can be recovered by organic solvent extraction of the biomass. Hydrocarbon recovery by mechanical squeezing is not efficient, since *B. braunii* has a thick cell wall and the wet biomass paste contains 10 times more water than hydrocarbons. Also, this recovery method cannot be applied to continuous culture of immobilized cells [16].

For organic solvent extraction, the solvent should have the following properties: readily available and inexpensive, immiscible with water, high hydrocarbon solubility, low boiling point for ease of removal, different density than water, and reusable. 9 solvents, which were compatible with *B. braunii*, were selected from 1500 compounds in [17]. Hexane was found to be the best to meet the above requirements and it was used as solvent for extracting hydrocarbons from *B. braunii*. For efficient extraction, the cells must first be concentrated to a paste because too much water presenting around the cells reduces contact of the nonpolar solvent with the outer cell wall where the hydrocarbons are stored [5]. Under suitable conditions, 70% of hydrocarbons were released by 30 min of contact with hexane [17]. Growth and hydrocarbon production are not affected by repeated extraction with hexane.

Culture immobilization via adsorption on polyurethane foams was used to improve the hydrocarbon recovery efficiency. Contact with hexane did not affect subsequent growth and hydrocarbon production of immobilized cultures. Recovery yields were markedly increased compared with free cells [18].

Microwave-assisted solvent extraction has been employed for the extraction of different polar and non-polar organic compounds from various solid matrices. Microwave irradiation is used as a heating source, and it has also been successfully used to assist leaching process [19]. Probably it can also be used in hydrocarbon extraction from microalgae with the great reduction of extraction time.

Scale up of extraction could be difficult, because the alga tends to aggregate and form clumps during extraction, and the polarities of the solvent and the wet cells are different. Clumping shields a large fraction of the biomass from exposure to solvent. Immobilization via adsorption on polyurethane foams was used to overcome this problem, and recovery yields were increased [18]. Agitation at high speed may also overcome this problem, but intense agitation damages the cells, however, it is quite expensive for large-scale commercial operations. Also, it is hard to scale up microwave-assisted solvent extraction.

6. Hydrocarbon Extraction by Supercritical CO_2

Recently green processes for the recovery of the hydrocarbons from biomass are gaining attention over the conventional techniques like steam distillation and liquid solvent extraction. Supercritical fluid extraction is considered as a green separation process for the extraction of natural products. The use of a non-toxic and volatile solvent, such as supercritical CO_2, which is inexpensive, easily removed from the extract, and reusable, protects extracts from thermal degradation and solvent contamination, leading to high quality products. Another advantage for supercritical CO_2 extraction is, it can be used as an efficient microalgae drying method.

Supercritical CO_2 has been applied to *B. braunii* for hydrocarbon extraction [16]. Figure 2 shows the diagram for supercritical CO_2 extraction setup. Because of its relatively low values of viscosity, density, and surface tension, supercritical CO_2 allows rapid extraction of solid materials. In the extraction of algal hydrocarbons with CO_2 at 30 °C and 40 °C, for different pressures of 12.5, 20, and 30 MPa, the highest rate of extraction was obtained at a pressure of 30 MPa [16].

The extraction yield and the fraction of the hydrocarbons in the extracts both increased with pressure, because the solubility of hydrocarbons increases with pressure. The CO_2 density increases with pressure, which implies a greater solvent power of the supercritical fluid. When the pressure increases, the dissolution of the outside hydrocarbons is easier than the intracellular lipids, and the rate-limiting step for mass transfer is the diffusion of these lipids through the cell wall, thus increase the hydrocarbon fraction in the extracts [20].

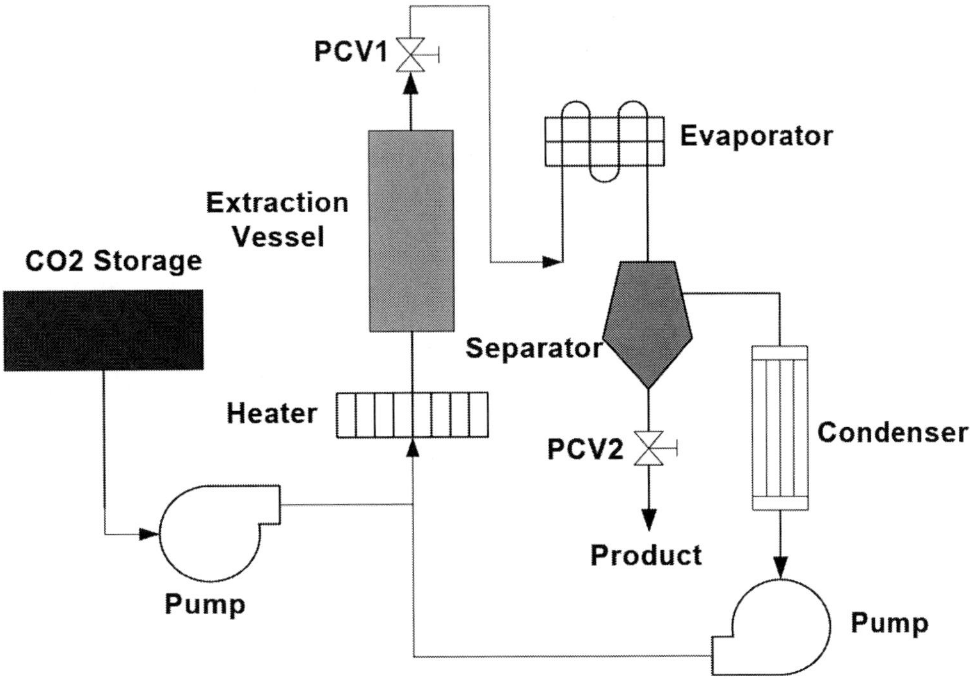

Figure 3. Diagram for supercritical CO_2 extraction

After supercritical CO_2 extraction, the extracts were fluid, golden and limpid, which was unlike those obtained from organic solvent extraction [21]. Chlorophyll and phospholipids were not detected. The residues of the alga after supercritical CO_2 extraction did not contain hydrocarbons; sterols were present in all the extracts and in the residue of the alga after extraction at 30 MPa most of the initial spot did not elute, showing that almost all the non-polar lipids were extracted [20]. The hydrocarbons of *B. braunii* were selectively extracted and this selectivity increased with pressure.

The progress of the supercritical extraction of hydrocarbons was as a function of CO_2 volume. The initial solubility of the hydrocarbons increased greatly with pressure. In the composition of the hydrocarbons in algal extracts, C27 and C29 hydrocarbons decreased proportionately during the progress of supercritical extraction, while the C31 hydrocarbon proportion increased. The lower molecular weight compounds were initially extracted preferentially, because their solubility in CO_2 is higher.

For other microalgae, similar results were obtained in literature [22]. Freeze-dried samples of the microalga *Chlorella vulgaris* were extracted by supercritical CO_2 at temperatures of 40 and 55°C and pressures up to 35.0 MPa. This study was carried out on whole and crushed algae. The extraction yields of carotenoids and other lipids increased with pressure, and the yields were higher for the crushed algae than that of whole alge.

Supercritical CO_2 extraction of carotenoids was more favorable with hexane and acetone extractions.

Although hydrocarbon extraction by supercritical CO_2 from microalgaes has advantages over organic solvent extraction, the drawback is, it is a semi-continuous process. Future applications will favor the continuous fractionation of liquid matrices with supercritical CO_2 in counter-current extraction columns, using much smaller volumes under pressure at much lower operating costs [23]. Nanofiltration membranes can be combined with supercritical CO_2 for this application, which will greatly facilitate the separation efficiency.

7. Direct Thermochemical Liquefaction of Microalgae

The hydrocarbon in *B. braunii* can be collected by extraction. Direct thermochemical liquefaction of wet and dry algal biomass can also be used to produce combustible oil. At 300°C and 10 MPa pressure, a good yield of oil was obtained in the presence of 5% Na_2CO_3 as a catalyst [23]. Thermochemical liquefaction of *B. braunii* eliminates the need for initial extraction step for recovery of hydrocarbons. However, the overall yield of oil from *B. braunii* was 52.9%, which is much lower than the extraction methods. A maximum recovery yield of 78% was achieved for botryococcenes in the liquefied oil at 200°C with the use of a catalyst [25]. The yields of the three fractions based on organics were 5% of lower molecular weight hydrocarbons (MW = 197–281, MW is molecular weight), 27.2% of botryococcenes (MW = 438–572) and 22.2% of polar substances (MW = 867–2209) [25].

Reference [26] studied the energy balance and CO_2 mitigating effect of a liquid fuel production process from *B. braunii* using thermochemical liquefaction. Thermochemical liquefaction has the advantage of treating wet materials compared with direct combustion, gasification and pyrolysis, because no drying process is required.

8. Converting Algal Hydrocarbons to Transportation Fuels

The hydrocarbons extracted from *B. braunii* are C23 and above with high boiling point, some of them are wax. They cannot be used directly as transportation fuel, thus further pyrolysis and catalytic cracking are needed.

Crude hydrocarbons of *B. braunii* can be catalytically cracked to 60-70% of gasoline, 10-15% of light cycle oil, 2-8% of heavy cycle oil, and 5-10% of coke [27]. Both Co-Mo catalyst and zeolite catalyst were successfully used in the crack. Yields of gasoline and other fractions varied with different catalysts and reaction temperatures. The optimal conversion temperature with zeolite catalyst was 497°C. The yields of gasoline obtained by catalytic cracking of algal hydrocarbons are comparable with that of petroleum. The gasoline has a sufficiently high octane number to be directly used as transportation fuel. Alkylbenzenes, mainly xylenes and trimethyl-benzenes in the product contribute to the high octane number. These aromatic hydrocarbons were derived from botryococcenes (the $C_{34}H_{58}$ compound) through consecutive cracking and cyclo-aromatization [5].

9. Summary

As one way to sequestrate CO_2 from flue gas emitted by coal-fired power plant, microalgae provide an economic solution for CO_2 mitigation as well as fuel production. Botryococcus braunii (*B. braunii*), which is a hydrocarbon-rich microalga, is selected because it is able to mitigate CO_2 and it has high hydrocarbon content as a source for renewable fuel production.

The hydrocarbons in *B. braunii* can be extracted by organic solvent or supercritical CO_2 from microalgae. Supercritical CO_2 extraction is an efficient method to extract the lipids and especially the hydrocarbons from microalgae and at higher pressures, it is more efficient for CO_2 extraction than by hexane extraction, due to the great increase of solubility of the hydrocarbons under pressure.

Fuel production from microalgae by direct thermochemical liquefaction eliminates the need for initial extraction step for recovery of hydrocarbons. However, the hydrocarbon recovery yield is lower than the method of extraction and cracking. Catalytic cracking of the hydrocarbons extracted from microalge by Co-Mo catalyst or zeolite catalyst provides a better solution: higher yield of hydrocarbon recovery and high quality fuel, which can be directly used as transportation fuel.

10. References

[1] T. Nakamura, C.L. Senior et al, April 2005, "Recovery and Sequestration of CO_2 from Stationary Combustion Systems by Photosynthesis of Microalgae", Final Report, DOE Contract No. DE-FC26-00NT40934.

[2] International Energy Agency (IEA), Carbon Dioxide Capture from Power Stations, 1998.

[3] Kiran L. Kadam, 1997, "Power plant flue gas as a source of CO_2 for microalgae cultivation: economic impact of different process options", *Energy Convers. Mgmt* Vol. 38, Suppl., pp. S505-S510.

[4] Masakazu Murakami and Masahiro Ikenouchi, 1997, "The biological CO_2 fixation and utilization project by rite (2) — Screening and breeding of microalgae with high capability in fixing CO_2", Energy Conversion and Management, Vol. 38, Supplement 1 (Proceedings of the Third International Conference on Carbon Dioxide Removal), pp. S493-S497.

[5] Anirban Banerjee, Rohit Sharma, Yusuf Chisti and U. C. Banerjee, 2002, "Botryococcus braunii: A Renewable Source of Hydrocarbons and Other Chemicals", Critical Reviews in Biotechnology, Vol. 22 (3), pp. 245-279.

[6] Min Gao, Bernd R.T. Simoneit, Miroslav Gantar, Rudolf Jaffe, 2007, "Occurrence and distribution of novel botryococcene hydrocarbons in freshwater wetlands of the Florida Everglades", Chemosphere Vol.70, pp. 224–236 and references therein.

[7] S. Sawayama, T. Minowa and S-Y. Yokoyama, 1999, "Possibility of renewable energy production and CO_2 mitigation by thermochemical liquefaction of microalgae", Biomass and Bioenergy, Vol. 17 (1), pp.33-39.

[8] Jian Qin, 2005, "Bio-Hydrocarbons from Algae: Impacts of temperature, light and salinity on algae growth", Rural Industries Research and Development Corporation, Australia, pp. 5.

[9] J. R. Maxwell, A. G. Douglas, G. Eglinton and A. McCormick, 1968, "The Botryococcenes—hydrocarbons of novel structure from the alga Botryococcus braunii, Kützing", Phytochemistry, Vol. 7 (12), pp. 2157-2171.

[10] A. C. Brown, B. A. Knights and Elsie Conway, 1969, "Hydrocarbon content and its relationship to physiological state in the green alga Botryococcus braunii", Phytochemistry, Vol. 8 (3), pp. 543-547.

[11] M. Neil Galbraith, Lance W. Hillen, Lindsay V. Wake, 1983, "Darwinene: A branched hydrocarbon from a green form of Botryococcus braunii", Phytochemistry, Vol. 22 (6), pp. 1441-1443.

[12] P. Metzger, E. Casadevall, M. J. Pouet and Y. Pouet, 1985, "Structures of some botryococcenes: branched hydrocarbons from the b-race of the green alga Botryococcus braunii", Phytochemistry, Vol. 24 (12), pp. 2995-3002.

[13] M. Murakami, H. Nakano, K. Yamaguchi, S. Konosu, O. Nakayama, Y. Matsumoto and H. Iwamoto, 1988, "Meijicoccene, a new cyclic hydrocarbon from *Botryococcus braunii*", Phytochemistry, Vol. 27 (2), pp. 455-457.

[14] L.W. Hillen et al, 1982, "Hydrocracking of the Oils of Botryococcus braunii to Transport Fuels". Biotechnology and Bioengineering, Vol. 24, pp.193-205.

[15] Eiichi Kojima and Kai Zhang, 1999, "Growth and hydrocarbon production of microalga *Botryococcus braunii* in bubble column photobioreactors", Journal of Bioscience and Bioengineering, Vol. 87 (6), pp. 811-815.

[16] Mendes, R. L., Coelho, J. P., Marrucho, I. J., Fernandes, H. L., Cabral, J. M. S., Novais, J. M., Palavra, A. F., 1995, "Applications of supercritical CO_2 extraction to microalgae and plants", J. Chem. Technol. Biotechnol., Vol. 62, 53-9.

[17] J. Frenz , C. Largeau , E. Casadevall, F. Kollerup , A. J. Daugulis, 1989, Hydrocarbon recovery and biocompatibility of solvents for extraction from cultures of *Botryococcus braunii*, Biotechnology and Bioengineering, Vol. 34 (6), pp.755.

[18] J. Frenz, C. Largeau and E. Casadevall, 1989, "Hydrocarbon recovery by extraction with a biocompatible solvent from free and immobilized cultures of Botryococcus braunii", Enzyme and Microbial Technology, Vol. 11 (11), pp.717-724.

[19] Mohamed El Hattab, Gerald Culioli, Louis Piovetti, Semch Eddine Chitour, Robert Valls, 2007, "Comparison of various extraction methods for identification and determination of volatile metabolites from the brown alga Dictyopteris membranacea", Journal of Chromatography A, Vol. 1143, pp.1–7.

[20] Mendes, R. L., Femandes, H. L., Coelho, J. P., Cabral, J. M. S., Palavra, A. F. & Novais, J. M., 1994, "Supercritical carbon dioxide extraction of hydrocarbons from the microalga *Botryococcus braunii", J. Appl. Phycol.,* Vol. 6, 289-93.

[21] Rui L. Mendes, Beatriz P. Nobre, Miguel T. Cardoso, Ana P. Pereira and António F. Palavra, 2003, "Supercritical carbon dioxide extraction of compounds with pharmaceutical importance from microalgae", Inorganic Chimica Acta, Vol. 356 (3), pp. 328-334.

[22] Rui L. Mendes, Helena L. Fernandes, Jose P. Coelbo, Elsa C. Reis, Joaquim M. S. Cabral, Julio M. Novaid, Anthio F. Palavra, 1995, "Supercritical CO_2, extraction of carotenuids and other lipids from *Chlorella vulgaris*", Food Chemistry, Vol. 53, pp. 99-103.

[23] A. V. M. Nunes, A. A. Matias, M. Nunes da Ponte, and C. M. M. Duarte, 2007, "Quaternary Phase Equilibria for scCO2 + Biophenolic Compound + Water + Ethanol", *J. Chem. Eng. Data*, Vol. *52,* 244-247.

[24] Shigeki Sawayama, Seiichi Inoue, Yutaka Dote and Shin-Ya Yokoyama, 1995, "CO_2 fixation and oil production through microalga", Energy Conversion and Management, Vol. 36 (6-9), pp.729-731.

[25] Seiichi Inoue, Yutaka Dote, Shigeki Sawayama, Tomoaki Minowa, Tomoko Ogi and Shin-ya Yokoyama, 1994, "Analysis of oil derived from liquefaction of Botryococcus Braunii", Biomass and Bioenergy, Vol. 6 (4), pp. 269-27.

[26] Yutaka Dote, Shigeki Sawayama, Seiichi Inoue, Tomoaki Minowa and Shin-ya Yokoyama, 1994, "Recovery of liquid fuel from hydrocarbon-rich microalgae by thermochemical liquefaction", Fuel, Vol. 73 (12), pp. 1855-1857.

[27] Kitazato, H., Asaoka, S., and Iwamoto, H., 1989, "Catalytic cracking of hydrocarbons from microalgae", *Sekiyu Gakkaishi*, Vol. 32, pp. 28.

THE PROPERTY VARIATIONS OF PRODUCED POLYACRYLAMIDE FROM POLYMER FLOODING RESERVOIRS AND THEIR EFFECTS ON CONCENTRATION DETECTION

Yefei Wang, China University of Petroleum, Dongying, Shangdong, China
Qing You, China University of Petroleum, Dongying, Shangdong, China
Lina Mu, China University of Petroleum, Dongying, Shangdong, China
Baojun Bai, Missouri University of Science & Technology, Rolla, MO, USA
Fulin Zhao, China University of Petroleum, Dongying, Shangdong, China

Abstract

Because the injected polymer (polyacrylamide) often quickly breaks through to production wells during polymer flooding, its concentration detection is very significant. However, due to the variations in the relative molecular mass and the degree of hydrolysis of produced polymer, the conventional detection methods, including the viscosimetry and starch-cadmium iodide colorimetry, are no longer suitable to detect its real concentration.

This paper introduced a new detection method-ultrafiltration concentration film's desaturation. By purifying and concentrating the produced polymer solution using an ultrafiltration system, then dehydrating the aqueous and dry up the remains, this method can detect the concentration directly. The said method has been evaluated in the laboratory and compared with the other two conventional detection methods, verifying its adaptability and high exactitude.

The produced polymers from three different polymer flooding units have been detected using this novel method. The results show that the produced polymer relative molecular mass decreased greatly but its degree of hydrolysis increased obviously. The reasons and differences have been analyzed. These corresponding variations makes the detected results by the conventional methods be much lower than the actual, meaning that the previous methods need to be improved or replaced.

Key words: Polymer Flooding; Produced effluents; Ultrafiltration, Polyacrylamide; Property, Detection

Introduction

The polymer flooding technology is currently the best established of the chemical enhanced oil recovery methods. Partially hydrolyzed polyacrylamide (HPAM) is primarily used commercial polymer. Its properties, usually referred to its average relative molecular mass and degree of hydrolysis, are closely bonding to flooding performance in the reservoir. Because the injected polymer often quickly breaks through to production wells during polymer flooding, it is very significant to monitor the produced polymer from the production wells. A simple, fast, and reliable analytical method for detecting produced polymer is especially helpful to understand the polymer flooding mechanism, guide the implementation of polymer flooding and comprehend the existing forms of polymer underground.

Compared with the ground injected polymer, the properties of the produced polymer vary greatly because the injected polymer has been subject to mutual effect of temperature, pressure, shear stresses, water-dissolved salt, oxygen, hydrogen sulfide, bacterial action as well as other chemical additives. The two property variations are degradation and hydrolysis of HPAM.[1, 2, 3]

Views of investigators concerning the problem of HPAM degradation in different conditions are quite contradictory, in most part based on speculative conclusions and often insufficiently confirmed by experimental data. But there exists convincing confirmations of HPAM degradation under the influence of thermal oxidation and bactericidal action.[4]

Thermal-oxidation is an important mechanism of degradation of HPAM; it appears that there exists a critical oxygen concentration for degradation of HPAM; a concentration level of 15 ppb is already too high to maintain its stability at high temperature (80-90℃).[5, 6]

The hydrolysis of HPAM is referred to the conversion of amide groups to carboxylic groups. Laboratory data showed that commercial HPAM hydrolyzed to an equilibrium degree that depends on the temperature, pH, initial carboxylic group content, but is largely independent of the brine composition and oxygen content.[7,8,9] It has been proved there is no variation of hydrolysis rate between pH 7 and 9, which can be explained by acid and base hydrolysis mechanism.[6]

The study on the long-termed chemical stability under simulated field conditions had been conducted by determining the degradation extent by comparing the current viscosity with initial viscosity, but the degree of hydrolysis does influence the viscosity too.[10, 11]

The produced polymer detection consists of polymer mass concentration monitoring and properties analysis. Due to the complexity of produced effluents, the detection has not been a simple problem because these samples also normally contain materials which show serious interference when using previously reported procedure. In the meantime, the property variations of produced polymer could not be achieved by a simply laboratory simulation because of the complexity of real reservoir conditions.

Currently, there are several methods to determine the concentration of injected polymer, including gel permeation chromatography (GPC), starch- cadmium iodide colorimetric, turbidometry, fluorescence spectrophotometry, viscosity comparison, sedimentometry, etc.[12,13,14] All these methods are usually used to analyze the polymer solution prepared by dry polymer powder. However, to the polymer in produced effluents from polymer flooding reservoirs, due to obvious variations of relative molecular mass and degree of hydrolysis of polymer, [15] the influences of impurities including brine and surfactants, there is doubt that whether these methods are suitable to determine the polymer concentration of the produced fluid or not. For example, both the viscosity comparison and starch-cadmium iodide colorimetric presently applied to determine the concentration in the oilfield all need to utilize calibration curves, which are always based on the property of dry HPAM powder.

So it is very significant to acquire the related data from a real reservoir. The only works which were conducted have been in Chateaurenard oilfield[16] whose reservoir temperature is very low, i.e.30℃. This paper is to make the further understanding about the produced polymer from different reservoir conditions, explore and discuss related problems.

Experimental Descriptions

Experimental methods

In order to analyze the produced HPAM, it is necessary to separate, purify and concentrate the produced effluent. The polymer containing effluents produced from production wells are treated as following steps: 1) Separate water phases from produced effluents and rest it, coarsely filter to get rid of free oil and solid suspensions; 2) Add reductant (such as iso-ascorbic acid) to deoxidize and stabilize the water samples; 3) Eliminate all emulsion oil and soluble oil using n-pentane extraction; 4) Filter using 5 μm Millipore filter and get rid of fine solid particles such as clays; 5) Ultrafilter the final samples by using the 100K or 300K BIOMAX PELLICON XL FILTER(products of Millipore Company) with a cut-off value equal to M_w = 100K or 300K Dalton respectively, and obtain the concentrated polymer solutions, whose metal ion concentrations were checked using ICPS to ensure their high purity so that they can be directly used to determine polymer mass concentrations by film desaturation or determine relative molecular mass and degree of hydrolysis by the conventional methods.

Experimental samples

Commercial HPAM products (in forms of dry powder) were from three different oilfields in China; Produced effluents containing HPAM were sampled from twelve production wells in the above-mentioned oilfields which were being conducted polymer flooding.

Results and Analyses

The validity of Method

Ultrafiltration has been widely used to eliminate low molecular mass impurities and concentrate macromolecules.[17, 18, 19] This paper used Millipore tangential fluid ultrafiltration system to lessen polymer retention on the membrane and improve filtration efficiency. In order to investigate the validity of ultrafiltration concentration film's desaturation method, the polymer solutions were prepared using the produced water which came from the different production wells, and detected the concentrations of these known solutions by the said method. The results showed that different mass concentrations, different polymer sources and different produced water had no effects on the detection of the mass concentration of polymer solution; the relative error was always lower than 1.0%.

The concentration of polymer with different relative molecular mass was also analyzed to inspect the applicability of the method. The process was as follows:

Polymer solutions with a concentration of 1000 $mg{\cdot}L^{-1}$ were prepared and sheared for different time by using high-velocity stirring to gain polymer with different relative molecular mass. The analysis results were shown in Table 1.

Table 1. Analysis Results of the Polymer with Different Relative Molecular Mass

Relative molecular mass	19.8×10^6	14.7×10^6	9.8×10^6	7.2×10^6	5.0×10^6	1.1×10^6
Known , $mg{\cdot}L^{-1}$	1000	1000	1000	1000	1000	1000
Measured , $mg{\cdot}L^{-1}$	1010	1015	1012	1001	999	1008
Relative error, %	1.0	1.5	1.2	0.1	-0.1	0.8

So it can be concluded that the method of ultrafiltration concentration film's desaturation was valid to determine the polymer mass concentration although the polymer relative molecular mass greatly changed.

Relative molecular mass and degree of hydrolysis of produced HPAM

Since pure HPAM solution had been acquired, the relative molecular mass would be figured out using intrinsic viscosity [η] which could be determined by viscometry, and the degree of hydrolysis of HPAM could be determined by acid titration. The results are shown in Table 2.

Table 2. Properties of HPAM

Oilfield name	Commercial HPAM		Produced HPAM		
	Relative molecular mass	Degree of hydrolysis , %	Production well	Relative molecular mass	Degree of hydrolysis , %
Gudao oilfield	17.12×10^6	22.3	6XN3 6X403 6-4 GD21-518 GD32N10 GD22-X529	4.55×10^6 3.95×10^6 3.32×10^6 4.65×10^6 7.68×10^6 3.88×10^6	43.0 28.7 32.9 39.2 40.9 50.6
Xinbei unit in Daqing oilfield	19.81×10^6	28.0	X4-44-P48 X4-44-P50 X4-43-P49	1.66×10^6 1.92×10^6 1.20×10^6	34.2 36.6 37.9
Shuanghe oilfield	15.23×10^6	23.7	T114 H140 J107	2.39×10^6 2.00×10^6 2.67×10^6	56.0 61.5 61.0

The experiments were also repeated by using PELLICON XL FILTER with 100K and 300K BIOMAX, the results were consistent with each other, indicating that the polymer with relative molecular mass under 300K was very rare.

From Table 2, the relative molecular mass of the produced HPAM decreased tremendously, the degree of hydrolysis increased obviously when compared with the commercial HPAM; and also there existed differences in different production wells and in different oilfields.

The main factors[20, 21] resulting in property variations of produced polymer included:

1) Degradations before polymer entering the formation. There were many factors causing degradation during this process, including mechanical degradation, chemical degradation and biodegradation, etc. And the degrees of different degradations were also different, but the shearing degradations during injection would be primary factor.

2) Degradations and hydrolysis during flow through the porous formations. Degradations would continue due to mutual factors such as physical and chemical reasons, but property variation of polymer would primarily be fast elevation in degree of hydrolysis because of a long-time stay under reservoir conditions.

3) Chromatographic separation due to the porous formations. There existed a complicated influence to produced polymer. Based on the fact of size distribution of polymer molecules, on the one hand, the larger polymer molecules tended to get to production wells slower than the smaller because the larger polymer molecules were more easily adsorbed onto the surfaces of the rocks or captured in the throats of the pores; on the other hand, the larger polymer molecules entered less capacity of reservoir pore volume due to their larger sizes, resulting in earlier arrival in production wells. So it would be seen that sampling date had also special effect on determination results.

The variation differences were also owed to different injection situations, reservoir conditions, flow conditions, well spacing, and retention time and so on.

Taking account of different reservoir conditions (Table 3), there would be a better understanding about above-mentioned variations of produced polymer.

Table 3. Reservoir Conditions of Different Oilfields

Oil field	Reservoir Conditions		
	Original permeability, μm^2	Temperature, °C	pH of reservoir water
Gudao oilfield	1.00	70	8.0
Xinbei unit in Daqing oilfield	0.482	50	7.6
Shuanghe oilfield	0.687	70	8.5

The high temperature and alkaline condition are more favorable to the hydrolysis, so the polymer degree of hydrolysis from Shuanghe oilfield was the highest; the largest relative molecular mass of the produced polymer from Gudao oilfield was related to reservoir's high permeability, developed channels.

Effects of property variations of produced polymer on the concentration detection

Effect of relative molecular mass variations

Using starch-cadmium iodide colorimetric method to determine the concentration of polymer with different relative molecular mass, it was found that the results were fairly consistent with the results gained by the said method, which demonstrates that the change of relative molecular mass of polymer has no effect on the determination of colormetric method. However, the deviation was very great to use viscosity comparison method to detect the concentration of the produced polymer. The reason was that the decrease of relative molecular mass brought about partly loss of polymer viscosifying ability, resulting in the phenomenon that the detected concentration was lower than the actual.

Effect of degree of hydrolysis variations

Both the said method and the colorimetric method were used to detect the concentration of polymer with different degree of hydrolysis. Table 4 showed the results. The polymer with different degree of hydrolysis was obtained by sodium hydroxide hydrolysis of HPAM.

Table 4. Effect of the Degree of Hydrolysis on the Detection Methods

HPAM degree of hydrolysis, %	HPAM concentration, $mg \cdot L^{-1}$		
	Known	Novel method	Colormetric method
28.0	1000	1010	1008
39.6	1000	1010	838
48.1	1000	1008	664
60.0	1000	1009	509

It was seen that, to a constant HPAM concentration, the concentration detected by the colorimetric method become smaller and smaller with the increase of degree of hydrolysis, while the novel method was not affected. It suggested that the colormetric method be no longer suitable to detect the concentration of produced polymer. The reason lay in that the increase of degree of hydrolysis meant the decrease of amount of the amide groups, which was exactly what colormetric method could detect.

Concentration Detection of produced polymer in field

Due to variations of relative molecular mass and degree of hydrolysis of produced polymer, the detection results would be inevitably wrong if the colormetric method or viscosity comparison method is still used. The laboratory results by the novel method and the field results by the conventional methods had been listed In Table 5. It could be obviously seen that the conventional methods needed to be improved or replaced.

Table 5. Comparison between Field and Laboratory Detection

Oil field	Production well	Detected concentration , $mg \cdot L^{-1}$	
		Field detection	Laboratory Detection
Gudao oilfield	6XN3	254	646
	6X403	330	748
	6-4	397	788
	GD21-518	457	627
	GD32N10	1232	1625
	GD22-X529	504	788
Xinbei unit in Daqing oilfield	X4-44-P48	360	706
	X4-44-P50	440	770
	X4-43-P49	460	811
Shuanghe oilfield	T114	470	687
	H140	470	731
	J107	382	602

Conclusions

1) The novel method about detection of produced HPAM from production wells - ultrafiltration concentration film desaturation - has been studied and established.

2) The experimental results indicate that relative molecular mass of the produced HPAM tremendously decreases and degree of hydrolysis obviously increases, when compared with those of injected HPAM. The related reasons have been analyzed.

3) Due to property variations of produced HPAM, its concentration detection method by using conventional coloarimetric or viscosity comparison method needs to be improved or replaced.

Acknowledgement

The authors wish to extend their sincere gratitude to Chang He, Zhiwu Zhang and Jiansheng Miao for their work providing the experimental samples from different oilfields.

References

1. Bailing Kong. "Study on the Hydrolysis of Polyacrylamide at High Temperature and Its Selection", Journal of Southwest Petroleum Institute, Chinese, 2000, 22(1):66-69.
2. Bailing Kong, Jiuming Luo. "Study on the Hydrolysis of Polyacrylamides at High Reservoir Temperature Conditions", Petroleum Exploration and Development, Chinese,1998, 25(6):67-69.
3. Bailing Kong. "Thermal Stability of Polyacrylamide Solution Prepared by Oilfield Produced Water", Petroleum Exploration and Development, Chinese, 2001, 28(1):66-67.
4. L.I.Tolstikh, N.I.Akimov, I.A.Golubeva, et al. "Degradation and Stabilization of Polyacrylamide in Polymer Flooding Conditions". International Journal of Polymeric Materials, 1992,

17:177-193.
5. J.H.Bae. "Effect of Oxygen on the Thermal Stability of Polymers and its Control in the Laboratory Tests", Polymeric Materials Science and Engineering, 1984, 51:97-101.
6. J. Francois, H.Kheradmand, V.Plazanet, et al. "Effect of Iron on Chemical Stability of Polyacrylamide", Polymeric Materials Science and Engineering, Proceedings of ACS Division of Polymeric Materials Science and Engineering, 1986, 55:694-699.
7. Ahmad Moradi-Araghi, H.Doe Peter. "Hydrolysis and Precipitation of Polyacrylamides in Hard Brines at Elevated Temperature". SPE Reservoir Engineering, 1987, (3):189-198.
8. E.N. Zilberman, et al. "High Temperature Hydrolysis of Polyacrylamides", Vysokomol. Soyed. 1979, B21:30-33.
9. M.Higuchi, R.Senju. "Kinetic Aspect of Alkaline Hydrolysis of Polyacrylamides", Polymer, 1972, 3:370-377.
10. S.H.Yang, L.E.Treiber. "Chemical Stability of Polyacrylamides under simulated field conditions", SPE Paper 14232, Presented at the 60th Annual Technical Conference and Exhibition of the Society of Petroleum Engineers held in Las Vegas, NV September 22-25, 1986.
11. R.G.Ryles. "Elevated Temperature Testing of Mobility Control Reagents", SPE Paper 12008, Presented at the 58th Annual Fall Technical Conference and Exhibition of the Society of Petroleum Engineers held in Dalas, TX, 1983.
12. Bailing Kong. "A Review on Methods for Determining HPAM Concentration of Aqueous Solutions", Oilfield Chemistry, Chinese, 1996, 13(3):284-288.
13. P.M.Beazley. "Quantitative Determination of Partially Hydrolyzed Polyacrylamide Polymers in Oil Field Production Water", Analytical Chemistry, 1985, 57(11):2098-2101.
14. Shiguang Yang, Lin Yang, Xiatong Rau. "Determination of Concentration of Polyacrylamide-An Improvement on Iodine-Starch Colorimetric Method", Journal of Southwest Petroleum Institute, Chinese, 1992, 14(2):105-110.
15. Yefei Wang, et al. "Determination Method of Relative Molecular Weight and Degree of Hydrolysis of Polyacrylamide in Produced Fluid from Polymer Flooding", Journal of the University of Petroleum, Chinese, 2006, 30 (1):90-92.
16. A.G.Putz, J.Lecourtier, Fr.Pau. "Monitoring Polymer Properties in Production Wells of Chateaurenard Oilfield", In Situ, 1991, 15(4):389-408.
17. Kllduff James, Jr.Walter J.Weber. "Transport and Separation of Organic Macromolecules in Ultrafiltration Processes", Envirn.Sci.Technol., 1992, 26:569-577.
18. Yu.I.Dytnerskii, Yu.K.Romanenko, R.G.Kocharov. "The Use of Ultrafiltration for the Purification of High-Molecular-Weight Compounds", Membr.Tekhnol.-Novoe Napravlenie Nauke Tekh., Russian, 1973:109-111.
19. Anthony R.Cooper, Robin G.Booth, Palo Alto Dynapol. "Purification of Polymeric Dyes by Ultrafiltration ", Journal of Applied Polymer Science, 1979, 23(5):1373-1384.
20. Yijiang Zhang, Fuli Cheng, Yajun Wang. "The Analysis of Systematic Degradation Factors during the Polymer Flooding", Oil &Gas field Surface Engineering, Chinese, 1998, 17(6):15-17.
21. Feng Huang, Hanxiang Fan, Zehua Dong, et al. "Study on Biodegradation of Partially Hydrolyzed Polyacrylamide by Sullfate Reducing Bacteria", Petroleum Processing and Petrochemicals, Chinese, 1999, 30(1):33-36.

95d Ultraviolet Analysis of the Toluene Fraction in Bitumen from Agbadu (Western Nigeria)

Theophilus K. Oladokun, Pure and Applied Chemistry, Ladoke Akintola University of Technology Ogbomoso., P.M.B. 4000 Ogbomoso Oyo state., Ogbomoso, 23438, Nigeria

Bitumen sample obtained from one of the wells sunk by the

Nigerian Bitumen Corporation (NBC 7) located opposite saint Stephen's primary school, Agbabu, Ondo state Nigeria was analyzed. The four fractions (saturates, simple aromatics, higher aromatics and polar compound) obtained from chromatographic analysis were further subjected to instrumental analysis of the Toluene fraction revealed the presence of some conjugated

bonds which could either be dienes or enols. Whether it is dienes or enols that are present, we cannot tell as only further analysis somewhat stronger than ultra violet can be able to tell us. This however is out of the scope of this analysis.

143f Kinetics of Heterogeneously Catalyzed Triglyceride Transesterification: A New Simulation Framework for Biodiesel Production

Mian Xu, Chemical Engg & Analytical Science, The University of Manchester, Manchester, United Kingdom and Jhuma Sadhukhan, Centre for Process Integration, Chemical Engineering & Analytical Science, The University of Manchester, Manchester, United Kingdom.

Biodiesel, an alternative diesel fuel, has become more attractive recently because of its environmental benefits and the increase in the petroleum price. Nowadays, most industrial applications of biodiesel production are performed by the transesterification of renewable biological sources based on homogeneous acid catalysts, which requires downstream neutralization and separation leading to a series of technical and environmental problems. However, heterogeneous catalyst can solve these issues, and be used as a better alternative for biodiesel production. Thus, a heuristic diffusion-reaction kinetic model has been established to simulate the transesterification of alkyl ester with methanol over a series of heterogeneous Cs-doped heteropolyacid catalysts. The novelty of this framework lies in detailed modeling of surface reacting kinetic phenomena and integrating that with particle-level transport phenomena all the way through to process design and optimisation, which has been done for biodiesel production process for the first time. This multi-disciplinary research combining chemistry, chemical engineering and process integration offers better insights into catalyst design and process intensification for the industrial application of Cs-doped heteropolyacid catalysts for biodiesel production. A case study of the transesterification of tributyrin with methanol has been demonstrated to establish the effectiveness of this methodology.

Keywords: biodiesel production, heterogeneous catalysts, internal diffusion, reaction kinetics

143g Development of Petrol-Alcohol-Water Microemulsion Fuel as a Substitute for Petrol and Ethanol-Blended Petrol

Manish Arora, Rajat Arora, Sandeep Mourya, **Amit A. Gaikwad**, K. G. Singh, Lalit Paliwal, R. K. Meena, Leena Mehendale, and Ashok N. Bhaskarwar. Department of Chemical Engineering, Indian Institute of Technology Delhi, Hauz Khas, New Delhi, India

In response to adverse environmental impacts from mass consumption of fossil petroleum fuels as well as international crude oil reserves being finite, most countries are attempting to find answers to meeting future demand for fuel for transport. Internationally, there is a growing acceptance that renewable ethanol fuel produced from biomass, with its associated environmental benefits, will be the transport fuel of choice for the future. Ethanol/ petroleum fuel blends directly address vehicle emissions and transport fuel security of supply issues. In addition to reducing currently regulated vehicle emissions, the renewable ethanol content of these fuels can result in a net reduction in the emission of unburnt hydrocarbons and particulate matter. Use of ethanol/ petroleum fuel blends initially in the existing vehicle fleet is essential to develop the technology and infrastructure necessary to support wide-scale production and use of ethanol fuel.

Petrol-alcohol-water based microemulsion fuel represents a new and potentially cost-effective option. Microemulsions are optically isotropic, transparent oil and water dispersions which are thermodynamically stable. Microemulsions have received considerable attention in recent years because of their interesting thermodynamic and physico-chemical behavior, and diverse applications. A lot of work has been done in areas like microemulsion gels and preparation of vitamin E emulsions. But preparation, characterization, and testing of microemulsion fuels, comprising of petrol and water, and especially of ethanol-blended petrol and water, remain virtually unexplored.

The petrol-alcohol-water based microemulsion fuel was prepared by stirring a blend of petrol, surfactant, co-surfactant, and water in definite proportions and it was produced under certain carefully defined conditions. To prepare a stable microemulsion fuel, a specific type of surfactant is required to be added in optimum quantity along with a co-surfactant, which in our case was ethanol (both anhydrous, and hydrous ethanol, i.e. ethanol obtained from sugar mills).

Experiments have been conducted to determine the minimum percentage of anhydrous ethanol that was needed to be added to the mixture of petrol and water, starting with fixed initial compositions of petrol (ranging from about 79%-99.5%)-and-water (ranging from about 0.5%-21%) mixtures. A fixed quantity of surfactant was added to this immiscible petrol-water mixture followed by a continuous addition of the co-surfactant (anhydrous ethanol) until a crystal clear mixture was obtained, i.e. a microemulsion (transparent and clear to the naked eye). Other sets of experiments were carried out by starting with fixed quantities of ethanol, surfactant, and petrol, followed by a continuous addition of the water to this mixture until it just about became turbid. A ternary phase diagram was plotted using these data. The diagram had ethanol and surfactant on one axis, water on another, and petrol on the third axis. A number of experiments were carried out to confirm the stability of all the compositions between the defined boundaries (which correspond to the stable microemulsion composition zone) by selecting some random compositions in between. Thus, the boundary compositions of petrol, ethanol, surfactant, water and the interior regions of composition were verified for the formation and stability of microemulsions.

Similar experiments were also carried out using ethanol obtained from sugar mills (hydrous ethanol) instead of anhydrous ethanol. In this case, fixed quantity of petrol, and surfactant was mixed in a 250 ml (air tight) sample bottle (Borosil). The resulting mixture was obviously turbid (unstable, two-phase region). Ethanol obtained from sugar mills (with different purities) was then added drop by drop with constant stirring, using a magnetic stirrer. The addition was continued till a clear transparent mixture was obtained indicating microemulsion formation. The procedure was repeated for different compositions of petrol, and surfactant.

Each microemulsion sample was kept separately in an airtight Borosil glass bottle and it was checked on everyday basis for any instability that could creep-in due to weather changes. Some of the samples were even kept in different environmental conditions to check for their stability. Use was made of an incubator to realize temperatures ranging from 0 °C to 40 °C. The microemulsions, which were stable on a particular day, did not necessarily imply that they would be stable afterwards. Some of them did become unstable (turbid) after some time, usually within days. The observed instability of microemulsions might have been due to the loss of co-surfactant (ethanol) by evaporation. Although the samples were kept in tightly closed sample bottles, the possibility of evaporation of the co-surfactant was still there. Another cause of instability of the microemulsion samples, occurring after some time, might be the change in the ambient temperature. In case a microemulsion sample turned unstable, it was stabilized using anhydrous ethanol and the composition of that particular microemulsion updated accordingly.

The results suggested that the formation of stable microemulsion started only at around 55 %v/v of ethanol. At high vol% of petrol (or low vol% of ethanol), a stable microemulsion was not obtained. The surfactant was quite effective in obtaining a stable microemulsion at a low volume percentage of about 0.29 – 0.33%. Using these experiments, we were successful in mapping the stable microemulsion zone on the ternary phase diagram for different compositions of petrol, anhydrous/hydrous ethanol, surfactant, and water.

INDALIN+: A New Process for Production of Auto LPG and High Quality Gasoline

Satyen Kumar Das*, Debasis Bhattacharyya, V. Krishnan

Indian Oil Corporation Limited, Research & Development Centre
Sector-13, Faridabad, Haryana, India
** email: dasak@iocrd.co.in*

Abstract

Under the emerging scenario of increasing stringent fuel specifications, some of the streams such as visbreaker / coker naphtha, straight run naphtha, etc. no longer qualify for blending into gasoline pool. In addition, the demand of straight run naphtha is declining due to its substitution by natural gas in fertilizer and power sectors.

On the other hand, the demand of LPG as cooking gas is in increasing trend in countries like India, China, Philippines, etc. LPG (rich in C3 & C4 paraffins) is emerging as a popular automobile fuel due to its several advantages. In such situation, a process for conversion of low value naphtha streams to LPG and high-octane gasoline is highly attractive proposition to the refiners.

In view of the above, Indian Oil R&D Centre has developed a new process 'INDALIN+' for conversion of naphtha / light gas oil streams of any origin including straight run naphtha and natural gas liquid to high yield of auto grade LPG and high quality gasoline and lower olefin content. The sulfur content in gasoline is significantly lower compared to the feed sulfur. The process can also be integrated with BTX plant for production of aromatics and further increase in LPG yield. INDALIN+ employs circulating fluidized bed reactor and regenerator hardware and proprietary catalyst formulation without requiring any external source of hydrogen and feed pre-treatment.

The paper discusses the Indalin+ technology in details. The papers outlines the wide spectrum of operating conditions, feed streams, catalysts can be used in the novel process. This paper also presents the different aspects of technology along with extensive pilot plant data.

Introduction

Under the emerging scenario of increasingly stringent specifications of gasoline in terms octane, olefin, sulfur and benzene content, it is difficult to include visbreaker naphtha, coker naphtha, etc. in the gasoline pool. Moreover, there has been significant increase in refining capacity and the fertilizer and power industries, the bulk consumers of naphtha, are gradually switching over to Natural Gas for obvious reasons. All these factors together are resulting in naphtha as surplus. Due to the similar reasons as above, the demand of Natural Gas Liquid (NGL) is also likely to reduce since it do not fit into the required fuel specifications. Therefore, upgradation of naphtha / NGL to higher value products seems to be essential to sustain the refinery operation in near future.

On the other hand, the demand of LPG is in increasing trend in India as well as other Asian countries, e.g., China, Philippines. The growth rate will further enhance if the use of LPG (rich in C_3 & C_4 paraffins) as automotive fuel in Indian metros catches up.

Keeping the above emerging issues in mind, Indian Oil R&D Centre has been working to develop a process for conversion of low value naphtha range stream including NGL to LPG and high-octane gasoline. Extensive micro-reactor and pilot plant studies have been conducted using various feeds, catalysts and fluidized reactor systems. This has resulted in development of INDALIN+ process. Besides producing higher LPG, the process also promotes simultaneous reduction of olefins and sulfur in liquid product. US patent for the process has been granted and application in India & Europe are pending. New patent applications have been filed for few more innovative steps. With the changing scenario of refining, use of tailor-made technology such as INDALIN+, will provide the much needed technology option for maximizing the value addition from surplus naphtha streams.

The present paper briefly outlines different aspects of INDALIN+ such as feed flexibility, novelty of the catalyst, process conditions, product yields and quality, application, etc. along typical utilities and economics of a commercial plant.

INDALIN+ process

'INDALIN+' is a new process for single step catalytic conversion of naphtha / light gas oil range hydrocarbon streams of any origin including NGL to high yield of LPG and premium quality gasoline. The salient features of the process are,

- High yield of LPG in the range of 35-65%
- Premium quality gasoline – higher octane, lower olefins & benzene
- Feed flexibility – can process feed of any origin in naphtha / kerosene/light gas oil range
- Flexibility to change the ratio of LPG/gasoline as per the demand
- No requirement of feed pre-treatment and external hydrogen supply
- High degree of desulfurization – lower sulfur content in liquid product
- Excellent integration with aromatics complex for production of petrochemical feedstock and increase the LPG production further

Process configuration

INDALIN+ employs circulating fluidized bed reactor and regenerator system similar to FCC. INDALIN+ plant consists of mainly two sections, (i) Reactor and regenerator system for carrying out the cracking reactions in presence of solid fluidizable micro-spherical catalysts and regeneration of the deactivated catalyst (Converter section) and (ii) Main fractionator column and gas concentration section for separation of reactor effluent stream into various products (MFGC section). The schematic diagram of the Converter section is shown in *Figure-1*.

The mixed naphtha stream after preheating is injected into a 'short riser', which is connected to a 'dense bed reactor' at the top to achieve the desired conversion. At the exit of the dense bed reactor, catalyst is allowed to enter into the 'stripper' and product effluent after separation of entrained catalyst is fed to MFGC section for separation into products of desired cut points. The stripped vapors are fed to the top of the reactor through an equalization line. The stripped catalyst then enters into the dense / turbulent bed 'regenerator'. Coke deposited on catalyst during reaction is burnt in presence of air to achieve low coke on regenerated catalyst. Flue gas after separation of entrained catalyst fines is passed through turbo-expander and / or waste heat boiler for energy recovery before venting through stack. The regenerated catalyst is circulated back to the bottom of the riser via a withdrawal well.

Catalyst & process conditions

Catalyst of INDALIN+ is micro-spherical, fluidizable solid particles having definite particle size distribution. It comprises several active ingredients having varying crystalline and amorphous acid sites in different proportions depending on feed and operating objectives. The physical properties of the catalyst components are almost similar and fall within the range of conventional FCC catalyst. Indian Oil R&D Centre has developed the formulation of this catalyst, which can be manufactured in India by third party based on its proprietary formulation.

The reactions are carried out at a temperature equivalent to reactor dense bed temperature of 470-510°C. To meet the heat demand in the riser-reactor and low coke on regenerated catalyst, INDALIN+ design considers a total combustion regenerator and thereby eliminates the need of CO boiler. The salient process conditions are given in *Table-1*. The relatively higher reactor pressure allows minimizing the wet gas compressor load and increases per pass conversion and LPG yield in the reactor.

Product yields & quality

LPG potential for different streams under INDALIN+ process conditions are shown in *Figure-2*. The LPG yield in this process is very high ranging between 35-65 wt% of feed depending on olefin content in feed. The product yields for different feeds are compared in *Table-2*. Feed-3 comprises 100% gas wellhead condensate having 95% ASTM D-86 point of 263°C. This indicates that the process is flexible and can accommodate even kerosene range streams.

The typical composition LPG obtained from INDALIN+ is shown in *Table-3* and it is found to be highly saturated (> 80 wt%) conforming to the specifications of auto-grade LPG. In order to reduce the benzene in gasoline, a splitter is provided where benzene rich light cut is separated. The properties of combined liquid product, benzene rich naphtha and benzene lean gasoline are summarized in *Table-4*. In

INDALIN+ process, olefins are not generated and rather the olefins in feed are converted resulting in significantvery low olefins content (about 2 wt%) in liquid product. Another important feature of INDALIN+ is the very high degree of desulfurization in the range of 80-95% depending on the feed. With feed sulfur of 3181 ppm, production of gasoline with only 158 ppm (*Table-4*) without any use of hydrogen is an outstanding achievement. Moreover, the gasoline octane is also very high (RON>92). All these features of INDALIN+ enable direct blending of gasoline product into the refinery MS pool if there is no limitation on aromatics content. The process has also adequate flexibility to change the ratio of LPG and gasoline yield to meet the seasonal variation in product demand.

Application of INDALIN+

INDALIN+ unit can be operated in two schemes, namely, 'gasoline' and 'aromatics'.

Gasoline scheme: Fraction boiling higher than gasoline, (if any, which depends on the feedstock) is separated and recycled to the riser to enhance the flexibility of meeting the coke demand. The benzene content in gasoline is achieved by providing a splitter as indicated above. The resultant gasoline can be directly included in the refinery MS pool.

Aromatics scheme: The liquid product of INDALIN+ is rich in aromatics comprising about 20% toluene, 15% xylenes and 7% benzene (wt/wt). In case gasoline is not possible to include into the refinery pool due to the lack of demand or limitation on part of aromatics content, the aromatics can be recovered in existing / planned BTX plant for petrochemical feedstocks. The raffinate from BTX plant is a good feedstock for INDALIN+, which therefore is recycled to the riser. This integration extinct the gasoline production and also increases the LPG yield. The final yield pattern under this scheme using Feed-2 (as indicated in *Table-2*) is compared in *Table-5*.

Utilities & economics

The typical figures for utilities and economics are summarized in Table-6. The capital cost estimate considers the installation in an existing refinery and thereby, does not include the cost towards land, site development, roads & buildings, office, township, etc. The payback period is quite attractive for INDALIN+.

Conclusion

INDALIN+ is a novel process for upgradation of naphtha range feedstocks of any origin including NGL. In addition to high yield of predominantly saturated LPG, the process produces a premium quality gasoline with significantly higher octane and lower olefins, benzene and sulfur content. Dense bed cracking coupled with higher reactor pressure and tailor made catalyst formulation favor various reactions simultaneously for achieving the above features of this process. INDALIN+ can be integrated with existing / planned BTX plant, for production of aromatics for use as petrochemical feedstocks and incremental LPG production. Due to moderate process conditions and also lower coke yield, the investment and operating cost are relatively lower making the economics of the process quite attractive. The process has immense potential for commercial exploitation to meet the large demand of LPG in India through conversion of low value surplus naphtha range streams.

Acknowledgement

The authors are grateful to IOCL management for permission to publish this work.

References

1. US Patent No. 5976356 (1999).
2. Mitchell B.R., Ind. Eng. Chem. Prod. Res. Dev. 19, 209 (1980).
3. US Patent No. 6656345 (2003).
4. US Patent No. 6455750 (2002).
5. US Patent No. 6307117 (2001).
6. US Patent No. 6548725 (2003).
7. "Advances in FCC catalyst/process/hardware Technology to Meet Challenges" Hydrocarbon Publishing Company, Vol. 1 & 2 (1997).
8. US patent No. 6222087(2001)
9. US patent No. 5171921(1992).
10. US patent No. 4350614 (1982).

Table-1: Typical process conditions of INDALIN+

Reactor dense bed temperature	^0C	470-510
Reactor pressure	kg/cm^2 (g)	4 - 8
Preheat temperature (blend naphtha)	^0C	325 - 425
WHSV in reactor	hr^{-1}	1-10
Catalyst to oil ratio (fresh feed basis)		3-8
Dense bed temperature	^0C	620 - 650
Coke on regenerated catalyst	wt%	< 0.05

Table-2: Product yield pattern of INDALIN+ with different feeds

Feed	**1**	**2**	**3**
Feed composition, wt%			
Saturates	87.3	74.6	64.8
Olefins	Nil	15.6	Nil
Aromatics	12.7	9.81	35.2
Product yields, wt%			
Dry gas	9.8	10.4	9
LPG	48.8	54.5	40
Light naphtha	5.8	4.8	5.69
Gasoline	33.5	28.2	33.43
LCO	-	-	9.78

Feed-1: predominantly paraffinic mix naphtha
Feed-2: 52% praffinic naphtha + 48% olefinic naphtha
Feed-3: NGL

Table-3: Typical composition of LPG

Components (wt% of LPG)	
Propane	56 – 69
Propylene	5 – 6
Isobutane	11 – 15
n-Butane	11 – 15
Isobutylene	1 - 3
1-Butene / t-2-Butene / c-2-Butene	2.5 -4.5

Table-4: Properties of light naphtha & gasoline

Properties	**Total liquid product**	**Light naphtha**	**Gasoline**
Benzene, vol%	6.58	38.8	0.5
Olefins, vol%	2.0	10.4	0.5
Total aromatics, vol%	56.4	38.8	58.4
Sulfur, ppm	193	394	158
RON	92	85	93

Note: Feed sulfur = 3181 ppm

Table-5: Product yields under different scheme of INDALIN+

Product yields, wt%	Gasoline scheme	Product yields, wt%	Aromatics scheme
Dry gas	10.4	Dry gas	11.7
LPG	54.5	LPG	60.9
Light naphtha / Gasoline	4.8 / 28.2	Aromatics	25.3

Table-6: Typical utilities & economics of INDALIN+

Utilities		Economics	
	Quantity		
HP steam (MT/hr)	14	Capital cost (± 30%) (Million US$)	113
MP steam (MT/hr)	10	Gross refinery margin	73
LP steam (MT/hr)	18	Net refinery margin	26
Cooling water (m3/hr)	2900	Pay back period (years)	3
Fuel oil (kg/hr)	2400	IRR (%)	25
Power (MWH)	8		

Note: Plant capacity: 1 MMTPA; Operating hours per year: 8000

Figure-1
Schematic diagram of converter section INDALIN+ plant

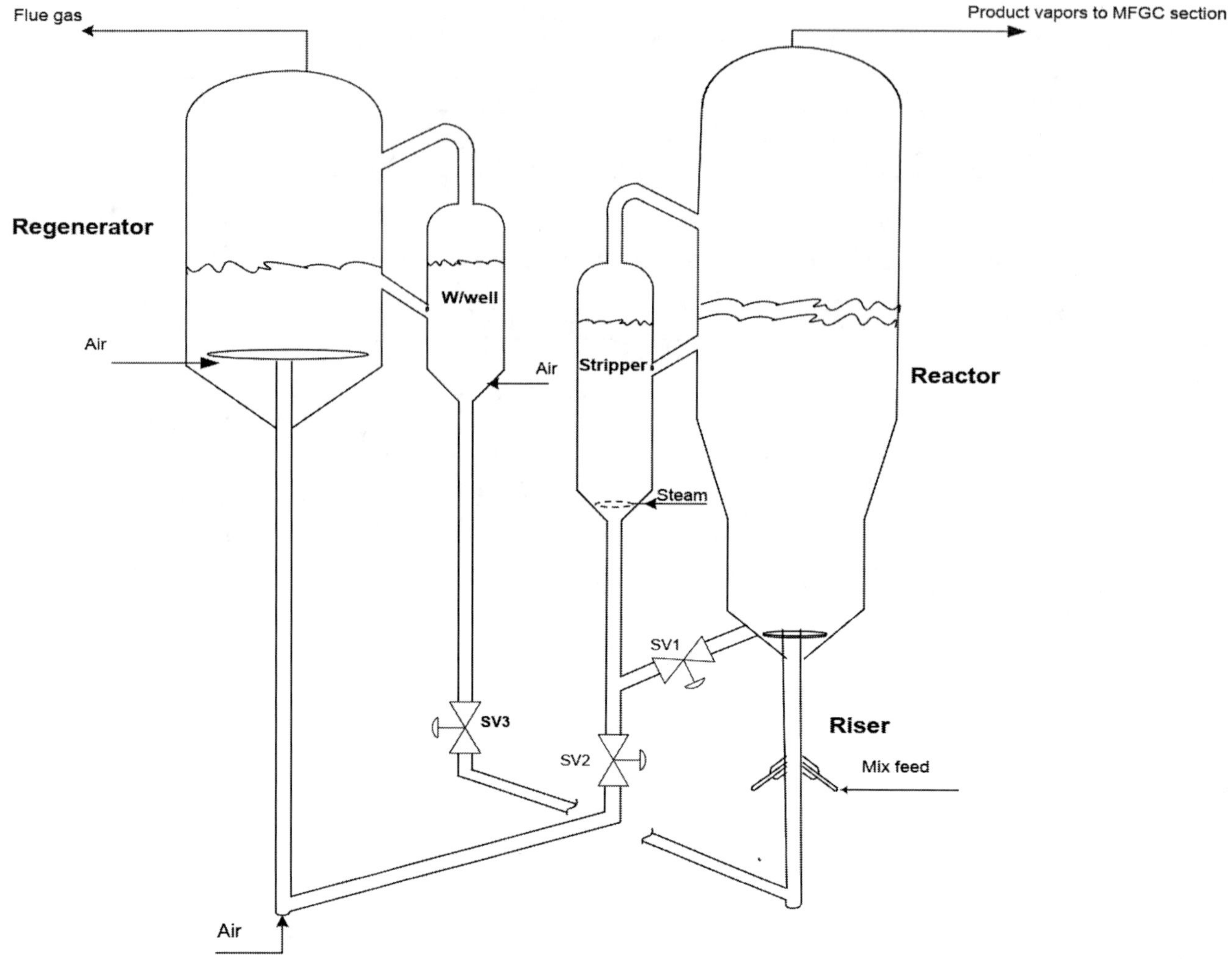

Figure-2
LPG potential (wt% of feed) of different feedstocks

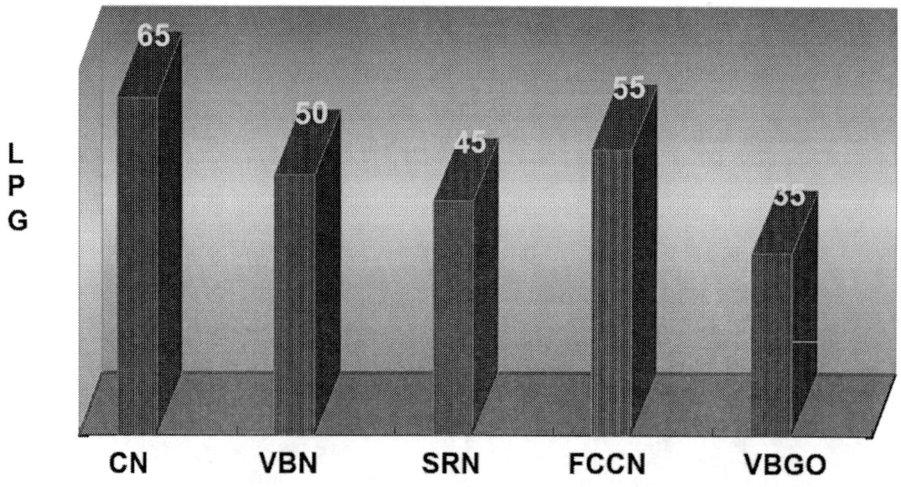

AUTHOR INDEX